CIM Justification and Optimisation

CIM Justification and Optimisation

Grier C. I. Lin & Sev V. Nagalingam

CRC Press
Taylor & Francis Group
Boca Raton London New York

CRC Press is an imprint of the
Taylor & Francis Group, an **informa** business
A TAYLOR & FRANCIS BOOK

First published 2000 by by Taylor & Francis

Published 2020 by CRC Press
Taylor & Francis Group
6000 Broken Sound Parkway NW, Suite 300
Boca Raton, FL 3487-2742

ISBN-13: 978-0-7484-0858-0 (hbk)

Publisher's Note

This book has been prepared from camera-ready copy provided by the authors.

British Library Cataloguing in Publication Data
A catalogue record for this book is available from the British Library

Library of Congress Cataloging in Publication Data
A catalogue record for this book has been requested

Contents

Preface

Manufacturing is one of the important pillars in sustaining the civilisation we are experiencing today. Almost all nations heavily rely on the manufacturing industry for their improved standard of living. On average, the manufacturing industry in the industrial nations and the newly industrialised countries contributes 30 to 60% of the gross domestic product, provides 35 to 45% of employment, and 25 to 60% of export earnings.

Integration plays a significant factor in maintaining a competitive and a capable manufacturing enterprise. Aspects of integration will dominate the beginning of the next millennium as well. Manufacturing managers and decision-makers in manufacturing enterprises have long understood the capabilities of individual technologies. However, two major hurdles remain towards the adoption of Advanced Manufacturing Technologies (AMTs) and Computer Integrated Manufacturing (CIM) in their enterprises. The barriers are the justification of AMTs in an integrated environment, and the utilisation of AMTs in an integrated environment to enable reaping the benefits of synergy

In today's ever changing global environment, to improve the well-being of all people, investing in advanced technologies in an intelligent manner to rejuvenate manufacturing enterprises is an important task. A need has been realised in order to address the concerns of manufacturing managers with respect to investment in AMTs and to guide them in a structured way to achieve and implement an integrated manufacturing environment such as CIM. This book is a result of eight years of research carried out at the Centre for Advanced Manufacturing Research, University of South Australia, by the authors, to address the concerns by the decision-makers in manufacturing enterprises.

The primary goal behind this book is to provide a justification and optimisation model that is complete enough to capture the important features required for a complex decision-making environment for CIM investment analysis, and is at the same time amenable for analysis. The model captures a company's strategic information, provides facilities to quantify qualitative attributes and analyses them alongside the quantitative attributes in a justification framework. Furthermore, the decision support model has the facility to enable investigation of the impact of subsystems, incorporate synergy among various functional units within a company and allocate resources in an optimal manner to satisfy the strategic considerations of an enterprise.

This book consists of nine chapters and two appendices in a simple readable form. A brief review of the developments in manufacturing technology and an explanation of what constitutes CIM are given in the first chapter. The second chapter reviews some of the available AMTs related to CIM, deals with issues on the importance of integration, provides information on enterprise modelling and presents a methodology for identifying functional inter-relationships in an enterprise. In the third chapter, developments made in the justification of CIM are explained. The chapter includes a brief introduction on traditional engineering economic tools and an explanation on the mechanisms behind highly utilised analytical techniques such as Analytic Hierarchy Process (AHP) and weighted scoring methods, which are used to evaluate strategic attributes. In addition, Chapter three introduces the reader to the shortcomings of traditional decision-making techniques, an insight to avoid the shortcomings and the Cost-Value technique as one of the effective value measuring techniques. The integrated approach to CIM justification and optimisation developed by the authors has been presented in three chapters for the logical understanding of the methodology. Chapter four describes the first part of the approach and helps identify the subsystems for the components of CIM, while satisfying the strategic requirements of the enterprise. Chapter five explains the second part of the approach. It consists of assessing the impact of the investments and mechanisms for justifying them. Chapter six describes an optimisation approach developed by the authors to allocate the resources of an enterprise in an optimal manner. Chapter seven describes the application of the methodology derived and describes the development of the decision support system, which is a key contribution of the research conducted by the authors. Chapter eight provides guidelines for implementing CIM systems. Finally, the conclusion is given as the ninth chapter and contains information on newly developing research directions on CIM for researchers in industry and academia. Two appendices are included at the close of the book. Appendix 1 describes some techniques which are necessary to apply AHP and Appendix 2 gives a summary of benefits for each AMT and the database used by the authors in illustrating the proposed methodology.

This book is aimed at providing sufficient information for readers to understand the implications of not investing in CIM, and the main principles and guidelines to invest and justify AMTs in a CIM environment. Present manufacturing managers will find this book a helpful tool that has information on justification and optimisation of CIM. In addition, it will be a useful textbook for future manufacturing managers (i.e. undergraduate and postgraduate manufacturing management and manufacturing engineering students), who strive to achieve confidence towards utilising AMTs.

Grier Lin and Sev Nagalingam
November 1999
Centre for Advanced Manufacturing Research
University of South Australia
Australia

CHAPTER ONE

What is CIM?

Why does this magnificent applied science which saves work and makes life easier brings us to little happiness? The simple answer runs: because we have not yet learned to make sensible use of it.

Albert Einstein

1.1 OVERVIEW

CIM is an acronym which stands for Computer Integrated Manufacturing. In today's parlance CIM has become a well-known term. This chapter reviews the historical developments in manufacturing and how CIM can be achieved. Before we go further into analysing CIM, looking backwards and realising how our manufacturing has grown as it is today will give an appreciation towards CIM.

1.2 WHAT IS MANUFACTURING?

Manufacturing is a process whereby products are made or created by various production activities from raw materials. The process of manufacturing enables us to enjoy our present way of life with many conveniences and tools. Manufacturing has evolved from simple procedures to the complex ones existing today. In addition, manufacturing can be regarded as the backbone of industrialised and newly developing nations of the world, since it provides more than one third of the economic activities of these nations.

The word manufacturing is derived from two Latin root words "hand" and "make". Early humans manufactured implements to help them carry out work with less manual power and almost all the tools and equipment were handmade, hence the word manufacturing was defined accordingly. The inventions and creations by humans have resulted in a variety of manufactured products. Many of the things surrounding us can be categorised as manufactured products. The clothes we wear; the communication tools we use everyday; the vehicles we use to travel are all manufactured products. It would be fitting to consider manufacturing as the pillar of civilisation, since manufacturing has changed the entire way of thinking. Human ingenuity continues to find ways to produce products which will improve the existing products and inevitably the betterment of the standard of living.

The industrial revolution which took place in the late eighteenth century introduced the concept of manufacturing and advanced the field of manufacturing with a special branch of engineering called manufacturing engineering. Manufacturing engineering gave birth to the concept of the factory as is known today by expanding the cottage industries, which were the main source of manufactured products. In the early days machines used manual power; later the power to drive these machines was produced by steam. The mechanisation of the manual procedures was the first step towards automation. This progress was mirrored in the usage of energy which originated from manual/animal power and advanced to thermal, petroleum and atomic power. The manufacturing process evolved from manual labour to mechanisation and from there to semi-automation and then to full-automation and today it is at the stage of computer integrated manufacturing (CIM). Utilisation of these new and innovative manufacturing technologies will result in improved economic activities and the prosperity of the nations, which encourage their enterprises to march forward with the help of manufacturing technologies.

1.2.1 Developments in Manufacturing Technology

The necessity to use automated machines has become an indispensable concept with the developments in technology and with the driving force of the demanding and sophisticated consumer market. Mechanisation facilitated mass production to meet the consumer demand for improved products. To achieve mass production transfer lines and fixed automation were created. Later, this further developed into programmable automation. The prime objective of the automation was to accelerate the production process throughout the plant. Automation has provided the ability to respond to the customers' demands while ensuring high quality products. Numerical Control (NC) was developed as an innovative approach to programmable automation around 1952 and gave a tremendous boost to the concept of mechanisation and automation. Later, NC evolved into Computer Numerical Control (CNC) and Direct Numerical Control (DNC) with the arrival of the computer technology. Computer Aided Design (CAD) paved the way for the application of computers in design and development. Combination of CAD and NC technologies resulted in Computer Aided Manufacturing (CAM).

With the developments in computer technology, the application of computers in manufacturing started to emerge, producing a wide variety of new technologies. Today there are many Advanced Manufacturing Technologies (AMTs) available off the shelf for the manufacturing companies to use and gain benefits of the automation. They include, office automation, CAD, CAM, Computer Aided Process Planning (CAPP), Computer Aided Quality Control (CAQC), Automated Storage and Retrieval Systems (AS/RS), robotics, Group Technology (GT), Cellular Manufacturing (CM), Flexible Manufacturing System (FMS), Computer Aided Engineering (CAE), Automated Guided Vehicle Systems (AGVS), Automated Assembly Systems, Automated Material Handling Systems (AMHS), Automatic Inspection, Statistical Process Control, Supplier Qualification Standards, Automatic Order Processing Systems, Integrated Production Scheduling Systems, Automated Shop Floor Data Collection Systems, and Computer Controlled Machines. These new and innovative technologies are either classified as AMTs or advanced manufacturing systems.

With the globalisation of the potential markets and production facilities, CIM has a further potential to propel towards a virtual CIM in today's perspective.

Virtual CIM describes a feature, whereby globally distributed CIM networks are interconnected through information and data sharing networks for the purpose of bringing together geographically dispersed enterprises to achieve the common goal of a business conglomerate. The historical developments made in manufacturing technology throughout the time can be described as in Figure 1.1.

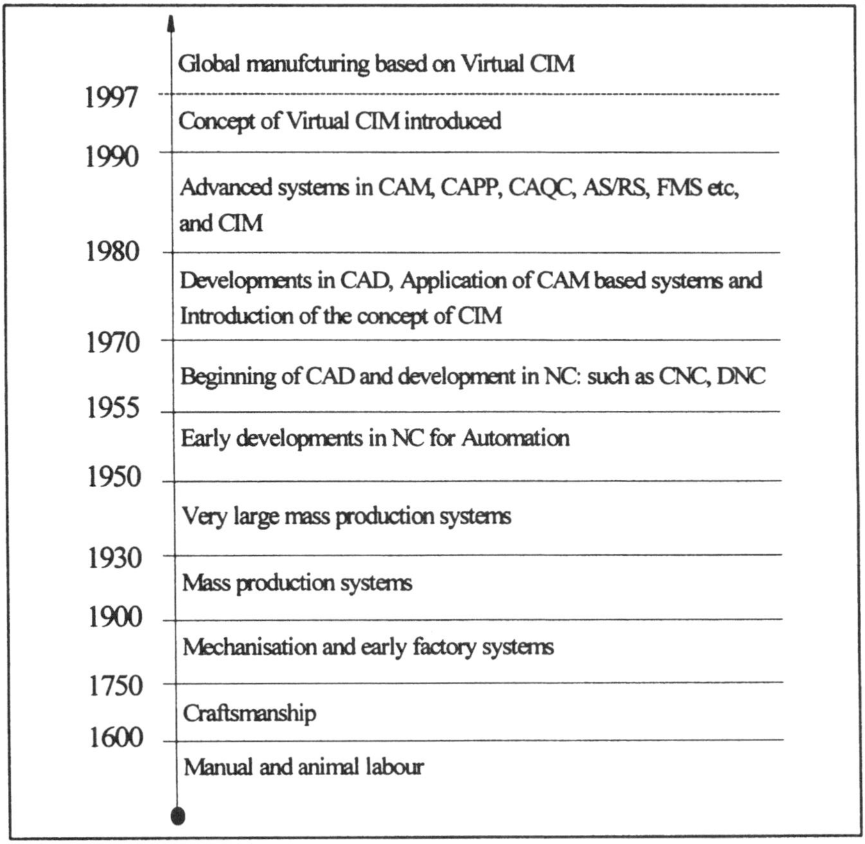

Figure 1.1 Developments in manufacturing technology

1.2.2 Present World Situation

The characteristics of the present world market (Figure 1.2, page 4) includes increased competition, shorter product life cycles, greater product diversity, variety and complexity, and smaller batch sizes to satisfy various customer requirements. Furthermore, non-price factors, such as product design, innovativeness, improved quality, enhanced features and delivery services are the primary determinants for the success of a product in today's global arena. To achieve these requirements, manufacturing companies need to be flexible, adaptable, responsive to changes, proactive and able to produce a variety of products in a short time at a low cost. In addition, they should be able to address new environmental requirements and concerns. Due to these needs, manufacturing companies are compelled to seek advanced technologies as an effective approach to solve the remedies of the existing manufacturing environment.

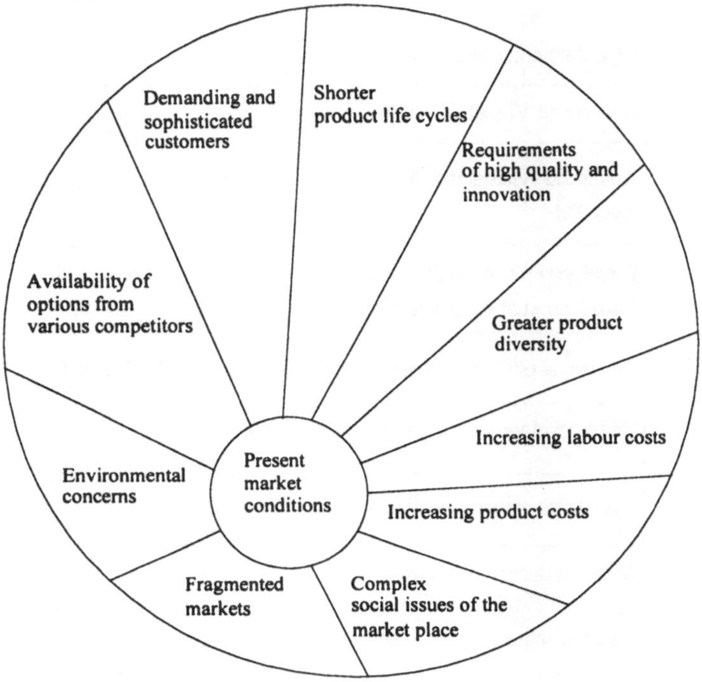

Figure 1.2 Present market conditions

The global economic competition, which exists today, forces manufacturing managers to consider and adopt innovative and advanced technologies. The manufacturing engineer must understand and be able to plan for these new technologies to survive in the highly competitive world. They should have a clear concept of automating the manual assembly and semi-automatic machinery to reap the benefits of these emerging technologies. Implementation of AMTs could help companies achieve their competitive goals to survive in the global market environment, as long as the technologies chosen are appropriate to meet their objectives. Having studied the productivity problems in 1976 Lin (1976) had stated, implementing integrated advanced technologies such as CIM, is an effective tool in solving the problems of decreased productivity relative to labour costs and consequent rise in unit costs which are continually plaguing present day manufacturing managers. Even after two decades Sohal (1997) reflected this view as implementation of AMTs provided opportunities to achieve the competitive advantage which is financially viable in an intermediate to long-term time frame.

1.2.3 Advanced Manufacturing Technologies

Advanced manufacturing technologies are generally defined as systems providing flexibility as well as data driven computer integration for a manufacturing organisation (Hunt, 1987). Advanced Manufacturing Technology (AMT) illustrates a technological application where the manufacturing technology utilised is intelligent enough to process the activities with less human intervention. In today's industrialised world, the enormous growth in manufacturing automation has brought a plethora of AMTs with diverse features. These AMTs could consist of semi- to fully-automated systems or equipment. Depending on the application of AMT and according to the level of integration, these AMTs can be categorised into four levels (Meredith, and Hill, 1987). The levels are:

1. Level 1 is defined as stand-alone systems which include NC machines, CNC machines, robots, CAD and others, or unitary equipment such as NC machine tools or robots. This type of equipment was generally controlled by self-contained computers or by programmable controllers;
2. Level 2 consists of computerised or non-computerised cellular groupings of equipment and materials for the production of families of parts; GT, FMS and CAE were classified under this level;
3. Level 3 mainly consists of linked islands of technology, which interconnects technologies in some functional units but the islands are regarded as separate entities without interconnection; and
4. Level 4 includes fully-integrated systems such as CIM, where design, planning execution of the planning for all aspects of the technology and the islands of automation in an integrated manner, links an entire company. This integration is carried out through a computer-controlled system or through automated information networks.

Whatever the classification is, AMTs are highly sought by the manufacturing companies to achieve a leading edge in the technology and to survive in highly competitive markets. The major objectives that motivate companies in the adoption of AMTs are as follows:

1. Enhancing efficiency and improving competitive position and profitability;
2. Enhancing production volume and achieving the flexibility in production;
3. Obtaining increased flexibility in responding to requested changes and demands;
4. Enhancing productivity and boosting the unit's image;
5. Optimal production flexibility and production scheduling;
6. Achieving more effective control of the manufacturing process;
7. Attaining higher quality in products and services;
8. Obtaining adequate and dynamic data for making decisions;
9. Overcoming the local skill deficiency and decreasing high labour costs;

10. Performing better inventory management and control;
11. Easing product/process design problems;
12. Performing improved environmental pollution control;
13. Increasing customers' dependence;
14. Preventing/minimising imitation of the application by others; and
15. Achieving increased market share.

1.2.4 Definitions of CIM

CIM is a process of using computers and communication networks to transform islands of enabling technologies into a highly interconnected manufacturing system. CIM involves integration of advanced technologies in various functional units of an enterprise in an effective manner to achieve the corporate objective of the manufacturing enterprise. There are numerous definitions available for CIM and the definition of CIM has had various connotations depending on the application of it. The Computer and Automated Systems Association (CASA) of the Society of Manufacturing Engineers defines CIM as an integrated system, which provides computer assistance to all business functions encompassing activities from accepting an order to delivery, within a manufacturing enterprise.

Whereas, the dictionary of Advanced Manufacturing Technology defines CIM as a concept of a totally automated factory, where all manufacturing processes are integrated and controlled by a computer (Hunt, 1987). IBM considers CIM as a broad based application, which provides a conceptual basis for integrating the application and information flow of product design, production planning and control and plant operations.

Although, CIM is regarded as an ultra-modern concept, it is actually a very old concept in a new form, formulated with the arrival of computer technology (Goetsch, 1988). In the early days the craftsman provided an integrated approach to manufacturing by single handily performing various tasks required to produce a product. However, with the technological advances all the tasks compelled specialisation in every activity and the manufacturing led to diversification and the demise of integration. The concept of CIM ties together all these tasks into an integrated manner through computer intelligence and networks again.

The prime purpose of CIM is to improve the overall profitability of the firm. This is accomplished by improving quality, productivity and flexibility through the integrated implementation of the constituent technologies of CIM and by achieving zero defects. A fully developed CIM system consists of integration, and application of each of the subsystems in the CIM environment in such a manner that the output of one system serves as an input to other systems.

1.3 BRIEF HISTORY OF CIM

Computers have been used in various applications for a long time. The first computer was created around 1946 and it became available to the public in 1951. In 1954, NC was introduced in manufacturing, and then in 1955 development of the first automatically programmed tool processor initiated the beginning of CAM. CAD began its appearance in the 1960s with high technology design conducted with computers. With the advent of the microchip in the beginning of the 1970s, the application of the computer has begun to enter all aspects of manufacturing.

The concept of CIM was formulated by Dr. Joseph Harrington in 1973 (Harrington, 1973). Despite his reluctance in formulating the abbreviation CIM, today the term CIM has become a widely known acronym in manufacturing parlance. Consequently the concept of CIM has been well entrenched into the management philosophy. CIM promulgates a fundamental strategy of integrating manufacturing facilities and systems in an enterprise through computers or intelligent processors and their peripherals.

A realistic long-term goal of CIM can only be achieved through company-wide planning and implementation. Effective integration requires an in-depth understanding of all the technologies and comprehensive knowledge of all activities in all functional units of a company. To be successful in the application of CIM, integration of AMTs should be achieved through the computers. The computers only act as subordinates to the technologies. However, without computers the integration cannot be effective and efficient since computers help organising, retrieving and storing information in an orderly manner with high accuracy and speed.

The department of defence in the USA played a vital role in pioneering CIM technology. In 1975 they started the Air Force Computer Aided Manufacturing (AFCAM) program, which was intended to provide a better approach for manufacturing technology. As a result of this, the ICAM program was initiated in 1976 at the materials laboratory of the USA Air force, under the guidance of the National Academy of Engineering Committee on Computer Aided Manufacturing. After discarding the ICAM program in 1985, the USA air force initiated the CIM program.

The Computer and Automated System Association of the Society of Manufacturing Engineers (CASA/SME) was founded in 1975 to provide a comprehensive and integrated focus in the field of computers and automation for the advancement of manufacturing. As an educational and scientific association CASA/SME spearheaded the concept of CIM to the general public and developed a CIM enterprise wheel to provide the manufacturing industry with a common vision of CIM. The technical council considered five fundamental dimensions such as: general business management; product and process definition; manufacturing planning and control; factory automation; and information resource management represented in the CIM wheel. The totality of the CIM enterprise wheel represents CIM as an enterprise-wide concept. It is shown in Figure 1.3 (page 8).

The CIM wheel is a top-down perspective of CIM and represents the integration aspect of CIM in the viewpoint of manufacturing management. CIM is an integration of islands of automation technologies in an effective manner. Nevertheless, compared to conventional manufacturing or islands of automation, if CIM is implemented the advantages are many fold. It brings:

1. Flexibility in product, volume and material;
2. Improved productivity and quality;
3. Improved interface between design and manufacturing;

4. Reduction in direct and indirect labour in various levels and functional units including middle management, engineering, administrative and operations support;
5. High quality designs which lead to good products;
6. High quality documentation and packages;
7. Better standardisation and better material utilisation;
8. Time and floor space savings;
9. A common database which eliminates huge data storage by independent systems;
10. Elimination of tedious, repetitive tasks and independent data storage;
11. Reduction in inspection time and work force associated with it; and
12. Competitive advantage over competitors.

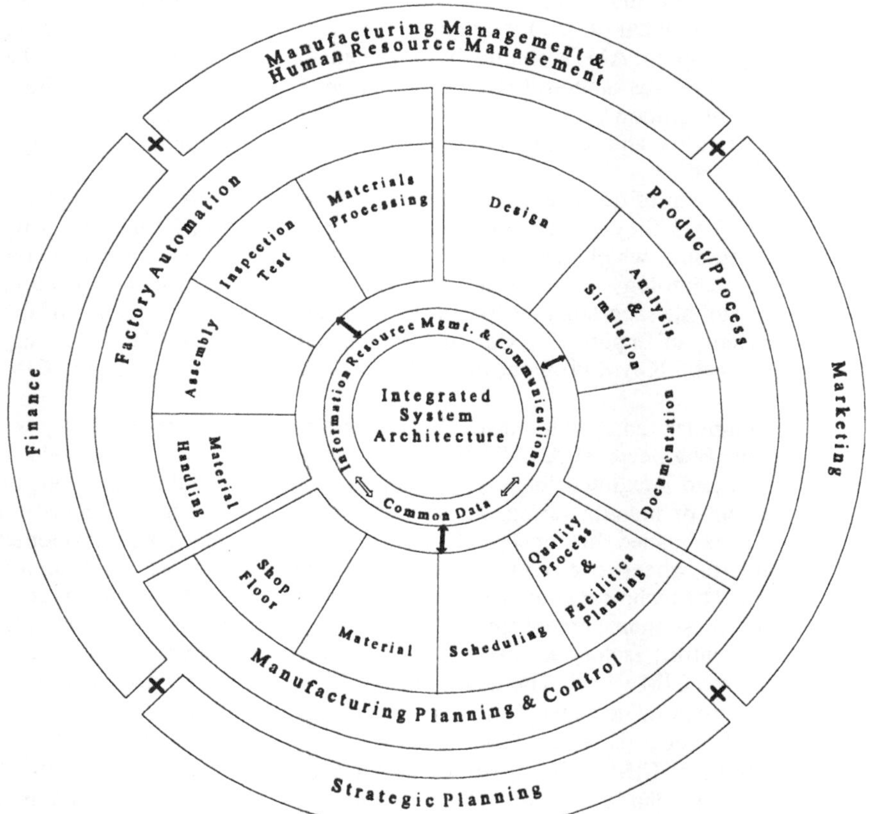

Figure 1.3 CASA/SME CIM Enterprise Wheel

The proposal of CIM in the early 70s may have appeared to be extremely futuristic, but in today's technologies it is possible to reach the goal of CIM without much difficulty. The future of engineering is inextricably bound with the application of CIM with embedded intelligence. Intelligence Manufacturing is the path for the future. Therefore, manufacturing technologies and CIM systems should be embedded with intelligence, which will help manufacturing enterprises to produce better quality products and make manufacturing equipment to solve problems encountered during the normal course of operation itself.

1.4 CIM MODELS AND CONCLUDING REMARKS

As outlined earlier, CIM results from the integration of many AMTs including the AMTs mentioned in this chapter (Section 1.2.1). Some of the most common AMTs are given as a brief illustration in Chapter two. However, CIM is not a turnkey system. It is a management and organisational approach which involves the integration of AMTs. A concept CIM model, which illustrates the concept of integrating technologies to achieve CIM, is shown in Figure 1.4 (page 10).

Activities are initiated by the customer order and all the functional units and the technologies in the organisation interact with each other in parallel and in sequence to achieve the order to delivery, with the help of an integrated database management system. Some of the CIM models adopted by IBM, National Institutes of Standards and Technology, Siemens and Digital Corporation are given by Rembold, Nnaji, and Storr (1994).

Even though there are many definitions for CIM and the models adopted by various companies can differ each in the level of integration and in conceptual level, the concept of CIM has become an important manufacturing requirement for success. However, the scope of the CIM model must include all functional units and the AMTs to achieve the objective of the enterprise.

Today many of the manufacturing enterprises are operating in islands of automation; CIM could be achieved by interconnecting these islands through enterprise integration and intelligent information systems. Nevertheless, justification of these interconnected technologies and an optimal connection is a paramount aspect of realising the benefits related to synergy, and to make a wise decision for the implementation of an effective integration.

To create a lasting employment opportunity for people and to improve the prosperity of the nation, we need to expand the existing manufacturing environment and prepare the decision-makers of the industries and government to face the challenges of the new millennium in a confident manner. Increasing the manufacturing base and taking advantage of innovative technologies and intelligent manufacturing will promote an environment of economic growth resulting in a positive trade balance for any country. To cater for this need new investment strategies and innovative industrial and technological approaches are necessary to be provided to the decision-makers. The aim of this book is to assist present and future manufacturing managers of the company to step in the right direction towards adopting AMTs and CIM to revitalise their enterprises.

This chapter, as the first step towards the stated goal, briefly reviewed developments in manufacturing and the concept of CIM. In addition, the present world situation will give an insight to the reader about the conditions faced by the manufactures to survive and to thrive successfully. Finally, a concept model of CIM is presented to illustrate CIM.

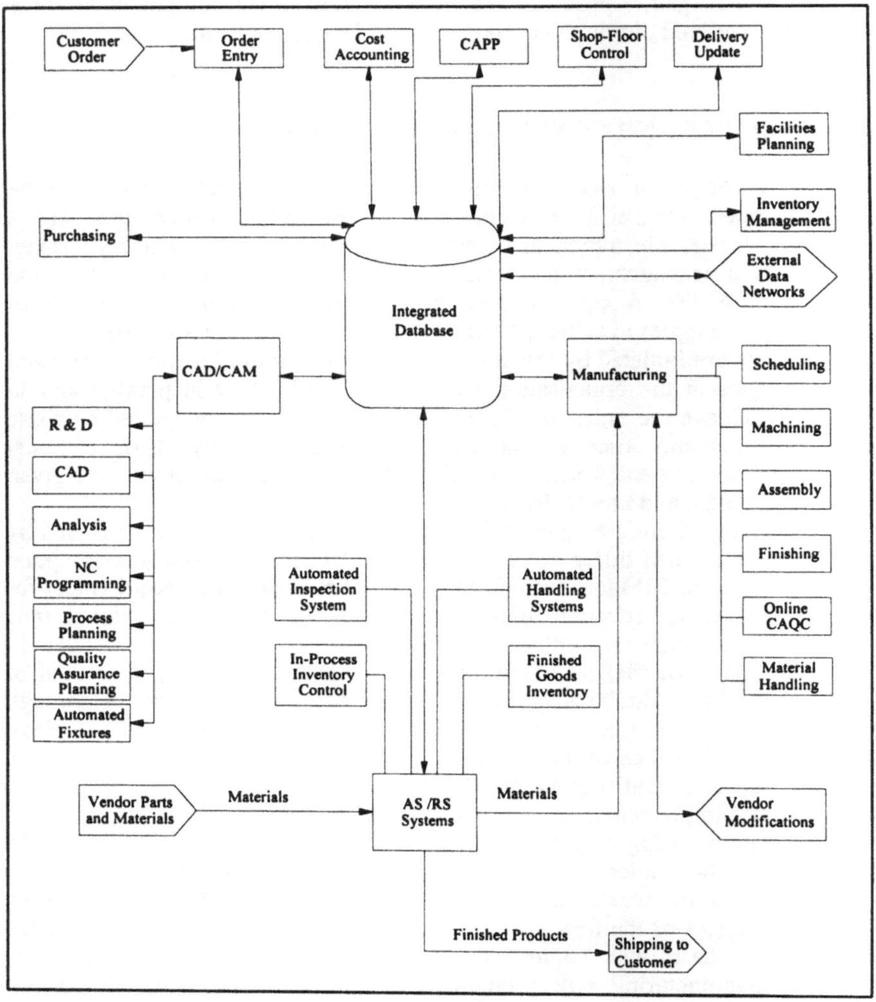

Figure 1.4 A concept of CIM model

1.5 REFERENCES

1. Goetsch, D.L., 1988, *Fundamentals of CIM Technology: automation in design drafting and manufacturing*, (Delmer Publishers).
2. Harrington, J. Jr. D., 1973, *Computer Integrated Manufacturing*, (New York: Industrial Press Inc)

3. Hunt, D.V., 1987, *Dictionary of Advanced Manufacturing Technology*, (New York: Elsevier Science).
4. Lin, G.C.I., 1976, "Integrated Computer Aided Design, Manufactuıe and Management", *Proceedings of Annual Engineering Conference*, Townsville, Australia, pp. 68–70.
5. Meredith, J.R. and Hill, M.M., 1987, "Justifying new manufacturing systems; a managerial approach", *Sloan Management Review*, Vol 28 No 4, pp. 49–61.
6. Rembold, U., Nnaji, B.O. and Storr, A., 1994, *Computer Integrated Manufacturing and Engineering*, (Addison-Wesley Publishing Company).
7. Sohal, A.S., 1997, "A longitudinal study of planning and implementation of advanced manufacturing technologies", *International Journal of Computer Integrated Manufacturing*, Vol 10 No 1–4, pp. 281–295.

CHAPTER TWO

Integration of Interconnected Networks

2.1 OVERVIEW

The objective of an integration program is to achieve data sharing between various islands of automation in an efficient and organised manner to improve the overall productivity of an enterprise. Manufacturing companies can invest in a wide range of AMTs to satisfy the customer requirements and the obligation to the shareholders. A clear understanding of the components of CIM is necessary for a successful implementation. Hence, in this chapter a brief introduction of the components of the integrated network, the architectures, and modelling techniques, which are necessary in integrating manufacturing enterprises and effectively eliminating the islands of automation, are given.

2.2 COMPONENTS OF CIM AND AVAILABLE AMTS

Computer Integrated Manufacturing as an emerging technology bringing new perspective to the management of companies via information sharing. Information technology is a powerful integration tool and part of the infrastructure that enables the desired behaviour in an integrated enterprise. In today's competitive world information plays a vital role in managing an enterprise. For several years researchers and consultants have been struggling with the problem of developing effective manufacturing information systems. Development of various communication standards including manufacturing automation protocol (MAP), technical office protocol (TOP), systems application architecture, initial graphics exchange standard, electronic design interchange format, business data interchange, ethernet, and token ring, had provided an impact in the implementation of integrated technologies such as CIM and opened a communication path for integration. Standardisation in data communications by protocols such as open system interconnection (OSI) facilitated an interconnection of possible diversified technologies and communication among them. In addition, interconnected information networks enable communication among managers, suppliers, employees and subsystems. Availability of information empowers managers and employees to fuse a strong and coherent corporate culture.

2.2.1 Integration of Subsystems

The technologies which can be connected in a CIM environment are numerous and may consist of various complex technologies. They may include technologies mentioned in Chapter one (Section 1.2) and many more. As developments in science and technology progress, many new advanced systems are becoming available for integration and manufacturing applications. However, in this book some of the AMTs are considered in an integrated environment to illustrate the concept in a simple manner (see Figure 2.1).

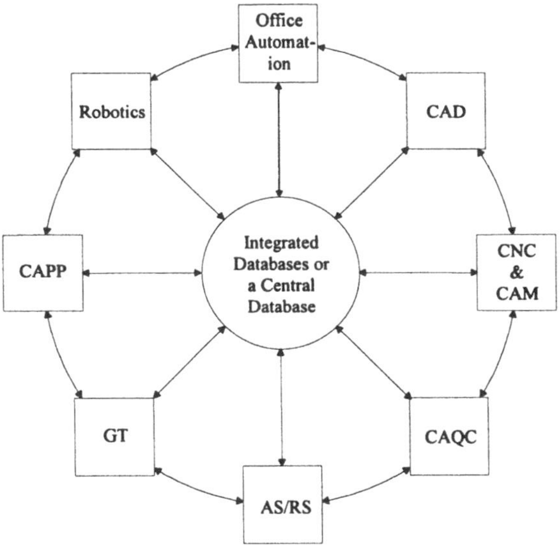

Figure 2.1 Integration of subsystems

Realisation of an effective integration by an efficient communication link among these advanced technologies will generate significant benefits across many traditional functional units and eliminate many functional barriers among the functional units. Similar to CIM, the AMTs also have many variations in their definitions. The definition of the term entirely depends on the level of complexity and technological sophistication of that technology. The definition of the term CAD can span from a simple computer aided drafting package to a complete computer aided design application which can generate simulation of the functionality of the object, analyse the object using finite element techniques, and generate the tool path for the numerical control machine. Similarly, the definition of the term CAM may vary from a simple computer controlled numerical machine for the application of the technology, to all manufacturing operations being controlled automatically including the manufacturing planning and control activities. Although providing definitions for the advanced technologies is a cumbersome task an attempt is made to provide a brief description about the AMTs considered in this book.

2.2.1.1 Office Automation

The general meaning of the term office automation can be considered as automating the office procedures by the available technologies. It may be considered as a simple computer terminal from where most of the office documents could be generated to a wide area network which has the facility to interlink all the manufacturing functional units and the technologies of the company by management information systems. This network could have a facility to capture all the requirements for the office such as generating the purchase orders directly to the suppliers by comparing the inventory level at the store and automatic sales order tracking. The automated office simplifies information and communication transfer in the form of data, written words, images, audio and video among various units. The application of computers in various factory management functions including production scheduling, material requirements planning, capacity requirements planning, cost estimating, process planning, purchasing, maintenance planning and inventory control, has provided many facets of office automation to the industry. However, it is crucial to establish the compatibility between various interconnected hardware and software of the office automation systems and the communication network.

2.2.1.2 Computer Aided Design

Computer Aided Design facilitates preparation of efficient design for manufacturing. CAD systems enable designers to optimise and improve their concepts of new products or products being developed by analysing numerous alternative configurations on the basis of many manufacturing priorities. CAD evolved from the results of research on interactive computer graphics in the late 1950s and early 1960s. SKETCHPAD, funded by the Department of Defence of USA and demonstrated in 1963 was a significant development in CAD. Many important developments in the area of computer technology in the 1960s and 1970s facilitated the development of CAD technology tremendously. Applications of CAD systems have been expanded to generate control procedures for CNC/ CAM and to produce process planning for CAPP. CAD will have a far greater cost impact on functional units such as marketing and manufacturing than it has in the drawing office. Since CAD can be used for tasks such as producing parts lists, quotations for estimating, simulating the machining process in the graphic screen, and generating machine codes for the manufacturing process rather than just producing drawings, in addition to the capability to produce more accurate drawings, application of CAD on these functions will have a significant effect on production costs.

For the incorporation of CAD into the CIM strategy, it is imperative that a CAD system has the capability to interface and access the computerised engineering database. The interaction of the engineering database with CAD can reduce product design and development time by reducing the time required to interpret design and establish production plans. Hence, it can be very helpful for the Research and Development (R & D) unit of the company as well. CAD can be used by R & D for many applications including the simulation and animation of the products, visualisation of the operational sequence, performance estimation of the product, finite element analysis and solid modelling. The effective use of a database depends on the ability of the CAD system to create libraries of commonly used component families and interpret parametric programs. Most of the present day CAD systems provide the facility of being able to communicate with other systems and having an interface to import/export data with other commonly used software products. This feature extends the application of CAD in manufacturing.

2.2.1.3 Computer Numerical Control

Numerical controlled machines are devices that enable machining of a workpiece according to the programmed instructions concerning the desired dimensions and the machining process for the workpiece. As explained in Section 1.2.1, the application of NC in industry can be regarded as the pioneer of AMTs. The control of machine tools using written instruction or coded programs is called NC, and NC encompasses a broad range of programmable machine tools. Prior to the use of NC all the machine tools were controlled manually by skilled operators. The quality of the product was very much dependent on the operator, whereas NC has eliminated the limitations of the human operators and improved the product quality.

The use of computers and programmable logic controllers coupled with the developments in manufacturing technologies pave the way for the CNC and DNC machines. CNC machines are provided with a monitor and keyboard so that writing and editing of the programs can be performed at the machine. In contrast, a DNC machine uses a mainframe computer or a minicomputer and many NC machines are simultaneously controlled. As with the developments in computer technology, DNC has evolved both in meaning and concept into a distributed numerical control, where each NC tool has microcomputer or programmable logic controllers, and all the systems are connected to a central host computer. These latest derivations of NC have solved many problems faced by NC technology and provided consistency in machining a set of parts in a batch type production environment. The use of NC technology enables the machinist to manufacture intricate shapes that are impossible by conventional machine tools. CNC machines are utilised in various manufacturing applications including machining, cleaning, finishing, inspection and quality control, and pressing and forming.

2.2.1.4 Computer Aided Manufacturing

Computer Aided Manufacturing provides the flexibility to be reprogrammed to perform a variety of manufacturing operations. Definitions for CAM systems are numerous. The use of computers in planning and control of a manufacturing facility through either direct or indirect computer interface with other physical, and logical resources of a company, could be regarded as CAM. CAM is expected to produce the machine codes for the NC machines from the drawing profiles created by a CAD system. A CAM system which has the simulation and the ability to link to other technologies such as CAD, NC, databases and quality control equipment, is essential for the success of CIM. CAM systems may consist of robotics, NC machines and FMS. However, in this book CAM is considered as an AMT which uses computer hardware and software to facilitate the automation and efficient production of industrial goods primarily on the factory floor by having a variety of machining capabilities.

2.2.1.5 Computer Aided Quality Control

Quality control is an important activity in a manufacturing enterprise. Quality control activities in a manufacturing environment consist of inspection and testing of a manufactured product. The inspection is carried out to verify whether the product conforms within the specific design standards. Highly sophisticated products with complex design features and a high level of manufacturing processes require an integrated automatic quality control facility. CAQC systems consist of computer-controlled Co-ordinate Measuring Machines (CMMs), laser beam devices and suitable software to carry out inspection and testing. With recent developments in the robotics technology, robots can also be used as a CAQC testing tool as well. Programmable CMMs are the widely used CAQC system. CMMs are utilised to gather data automatically from the engineering design database and to determine whether a component had been manufactured within the design tolerances specified in the design. Testing of the product verifies whether the product confirms to the functional aspect of the product. Depending on the requirement, various non-destructive and destructive testing procedures can be conducted to test the functionality of a product. CAQC improves the quality and the productivity of the company by the automated testing and inspection processes through the application of the computer technology with an advanced sensor technology.

2.2.1.6 Automatic Storage and Retrieval Systems

Automatic Storage and Retrieval Systems (AS/RS) are computer operated material handling systems, which deliver and retrieve a batch of parts to and from a specified location or to and from a random location, while recording the locations for future reference. Some Automated Material Handling Systems (AMHSs) are generally employed to move items or workpieces between workstations and specific areas. There are many variations in the type used in a manufacturing environment including assembly line vehicles, automated carriers that follow wires embedded in the floor, conveyors, light load vehicles, monorails, motorised carts on tracks, pallet trucks, towing vehicles, and unit load vehicles. An AGVS also is a form of an AMHS. An AS/RS can be considered as a controllable system which may facilitate handling, storing and retrieving of materials with precision and speed under a defined degree of automation. In this book the AS/RS is construed as an AMHS with the combination of the conventional AS/RS which could handle and shuttle workpieces between various workstations in a manufacturing environment in addition to having the capacity to store and retrieve while recording the present location for future use. The AS/RS should be able to identify an item from one of the following requirements or with a minimum number of combinations: the item's code number or any similar identity, the material of the item, the minimum number of dimensions of the item, and its location parameters.

2.2.1.7 Group Technology

Group Technology is a manufacturing concept where the parts to be produced are grouped according to their design characteristics or the manufacturing processes required to produce them. Parts which are grouped according to their design characteristics are classified as a design part family, and parts grouped according to their manufacturing characteristics are classified as a manufacturing part family. The grouping of parts by these characteristics to solve similar problems in manufacturing offers an improved efficiency in producing the desired components. An application of GT consists of the collection of machines arranged to produce a specific group of components according to their feature. Thus, the GT facilitates a system approach for the implementation of FMS or CM systems at the organisation. GT offers extensive benefits to a company which adopts it including improved design, reduced material handling, simplified production scheduling, improved quality control, an ability to react quickly to changing market conditions, reduction in scrap and rework and better planning facilities.

2.2.1.8 Computer Aided Process Planning

The use of computers for the preparation of process routing or process planning can be collectively defined as CAPP. Process planning is responsible for the creation of detailed planning instructions for a production of a part or for an assembly operation. CAPP is an interactive system which allows the process planner to generate process plans for parts, modify existing plans to suite the new parts, or carry out any other planning operations very quickly and easily. In addition, CAPP helps production planners to identify the optimal sequence of production operations for a product. The developed process plan for a component should be consistent with the quality standards and appropriate procedures for that component. CAPP systems are designed according to two approaches.

The first and most popularly based approach, the retrieval CAPP system (or variant CAPP system) is based on the principles of GT and its parts classification. For this retrieval approach a search is made for each part among existing standard process plans which are stored as computer files for various parts in a database. If the file contains a process plan for the required part, it is retrieved and displayed for the human planner for modification or approval, otherwise the process plan is generated by editing a process plan similar to the required part.

The second approach, which is a generative CAPP system is based on a concept similar to an expert system. This approach consists of automatic generation of process plans with the use of pre-defined algorithms which simulates logical procedures of the human planner, manufacturing standards and procedures from the knowledge-base. The knowledge-base could contain all the relevant machining information including feeds, speeds, finishing tolerances and power requirements for a required part in addition to the process procedures of the part. The generative CAPP system assists in analysing process planning systematically while producing optimal plans compared to the variant models. Greater benefits could be achieved by integrating the CAPP systems with other AMTs in the company and sharing a common database.

2.2.1.9 Cellular Manufacturing

The use of specified production areas called manufacturing cells for the manufacture of part families is called CM. Application of GT in a manufacturing procedure gave birth to CM. In CM, the manufacturing system of the company is organised into manufacturing cells. These manufacturing cells consist of machinery and processes closely located to manufacture a specified part family. Traditionally the layout of the manufacturing system of a company consisted of similar machines arranged together. This arrangement results in much of the manufacturing time spent on moving the object or waiting for a process, hence resulting in long lead-times. Grouping of the necessary machinery and processes together in a cell for the purpose of manufacturing a specified part family reduces the lead-time dramatically. The CM could be beneficial to a company when the production requirement of the company falls between jobs which are wide in variety and low in volume and jobs which are high in volume and require dedicated mass production lines.

2.2.1.10 Robots

Robots are programmable manipulators that can be used to move objects, workpieces or tools along various paths depending on the requirement. The definition of the Robot varies depending on what constitutes a robot. The Robot Institute of America (RIA) defines a robot as a "reprogrammable multifunctional manipulator designed to move material, parts, tools, or specialised devices through variable programmed motions for the performance of the a variety of tasks" (Goetsch, 1988). In contrast, the Japan Industrial Robot Association considers almost any machine that can be used to manipulate objects (Hunt, 1987). In this book the definition of RIA is construed for robots.

Industrial robots became a reality by the combination of the technological developments in the industrial automation technology, which has a long history of progress, and from the recent developments in computer science and artificial intelligence. Although the robots that fit within the definition came in to existence within the last three decades, the first articulated arm which utilised an NC technology and remote manipulation technology was used by the US Atomic Energy Commission in 1951. Today robots are used in a variety of manufacturing applications including material handling, loading and unloading machine components, painting, welding, machining, assembly, inspection, and sophisticated manufacturing procedures. Many of the initial applications of robots have been committed for particularly unpleasant and dangerous tasks. However, many manufacturing managers generally regard robots as a machine that replaces people, and it has been justified on that basis in the past. Considering the robot as a replacement of human labour limits many opportunities while creating many labour problems, whereas applications of the robot can give the establishment a leading edge in manufacturing operations. When robots are sought for the applications of unpleasant tasks, where the complexity of the job and very high accuracy is a priority, or when the application is in hazardous circumstances, the evaluation of the robot on the financial criteria makes it impossible to justify. Hence, a proper justification of robots should incorporate all the benefits by tangible and intangible attributes and not limit the evaluation on the financial considerations alone.

2.2.1.11 Flexible Manufacturing System

Flexibility is the capability to respond to new situations easily and quickly. However, the concept of flexibility is relative rather than absolute and has been defined by various authors in various forms. FMS can be considered as a group of processing stations interconnected by an automated material handling and storage system and controlled by computer systems. Processing stations may consist of a variety of manufacturing facilities including assembly stations, forging stations, metal presses, and inspection stations depending on the need. The concept of FMS revolves around the ability of the manufacturing system to respond to new products and changes of volume easily and quickly. This ability is obtained by processing various components simultaneously by interconnecting various technologies into a single production system under the control of a common computer. The FMS could be regarded as the beginner version for CIM. Applications of FMS could be considered in an enterprise where the production demands fall between high production transfer lines and low volume stand-alone NC machines, while providing the opportunity to meet demands for flexibility, quality and cost effectiveness simultaneously. In the specific case of FMS, flexibility can be considered to consist of the following forms (Tidd, 1991):

1. Expansion flexibility: The capability of the system to expand significantly and modularly;
2. Machine flexibility: The ability and the simplicity to make changes required to produce given products;
3. Operation flexibility: The capability to interchange the order of the productions operations for a product or to a product type;
4. Process flexibility: The capability to change the process of producing a given product with different materials and procedures with ease;
5. Product flexibility: The capability to change quickly, easily and economically for the production of new product;
6. Production flexibility: The ability to produce a different set of product under different circumstances;
7. Routing flexibility: The ability to continue production via alternative procedures in the case of breakdowns and the capability to cope in that situation; and
8. Volume flexibility: The capability to produce a different set of volumes profitably and the ability to switch over quickly and easily.

In reality, a particular FMS may not be able to provide all the forms of flexibility the company requires, hence the manufacturing manager should be aware of the combinations of flexibility the company requires, and have the capability to make an appropriate decision.

2.3 THE NEED FOR INTEGRATION

Traditionally, manufacturing companies install equipment whenever a need arises for it as a stand-alone unit. Later, when the demand for automation arises they attempt to interconnect technologies together. The need for integration has evolved in response to the problems faced by the traditional manufacturing process of industrial automation. Individual automation in each functional unit has created islands of automation. These islands of automation do not facilitate the communication between functional units. Errors in data sharing and other mismatches with these islands of automation continually plagued the manufacturing industry. The complexity of new manufacturing technologies, economics, increasing human limitations, computer developments, and competition from abroad has forced the initiation of the Integrated Computer Aided Manufacturing (ICAM) program in late 1970s by the USA Air force. The program lasted approximately eight years and concluded in the early 1980s. The ICAM program conducted a study in 1983 and found the following critical problems in industrial automation (Appleton, 1985):

- Information could not be controlled by users;
- Changes were too costly and time consuming;
- Systems were not integrated; and
- Data quality was not suitable for integration.

These problems were the result of the job shop approach to automation emanating from the growth of the traditional industries. When companies progressed from the job shop approach they were having to automated facility, they created islands of automation. When properly planned integration was not carried out, and only islands of automation were implemented, the completed benefits of automation were not obtained. Further, the lack of planning in integrating these islands created difficulties towards effective integration and improved only the local productivity, but they were insufficient to provide necessary logistic support to improve the productivity and quality throughout the company. To overcome these issues the ICAM program was established. The ICAM program was a practical effort to provide compatible techniques for integration and a unified direction to industry towards integrating technologies.

The prime element in an integration concept is having a common database, by which the data can be transferred automatically among various units and user groups. By introducing an integrated manufacturing concept, simplified workflow and reduction in the accumulation of work in progress will be the standard beneficial feature. The integration of the technologies brings following benefits:

1. Creation of a truly interactive system which enables manufacturing functions to communicate easily with other relevant functional units;
2. Accurate data transferability among the manufacturing plant or subcontracting facilities at in-plant or diverse locations;
3. Faster responses to data-changes for manufacturing flexibility;
4. Increased flexibility towards introduction of new products;
5. Improved accuracy and quality in the manufacturing process;

6. Improved quality of the products;
7. Control of data-flow among various units and maintenance of user library files for system-wide data;
8. Reduction of lead times which generates a competitive advantage; and
9. Streamlined manufacturing flow from order to delivery.

Although the benefits of integrated technologies are very difficult to quantify, integration provides a competitive advantage by linking new and existing hardware and software of the technologies, together with database management systems and data communications systems into a co-ordinated and efficiently managed process. Voss (1988) has stressed the importance of integration by stating that the benefits of the new manufacturing technologies will be greater with cross-functional applications than with unlinked functional applications. Many additional tangible and intangible benefits can be obtained by considering cross-functional approach and integrating various technologies across functional units.

The benefits attained with respect to integration of AMTs in an enterprise have been highlighted by many researchers and consultants. Before integration can be implemented, an organisation must understand its existing activities and how these activities will be effective under an integrated framework. Lack of management understanding on how the integration across functional units can benefit an organisation is a hindrance in adopting CIM. Furthermore, implementation of CIM requires an appreciation of AMTs and the vision to foresee the benefits of the integrated technologies to an enterprise. Managers and the employees need to have a clear vision, appropriate strategies, and goals of the organisation for the future to achieve success through integration in an integrated enterprise.

The benefits of integrated AMTs are not limited only to the functional unit where the technology has been implemented. The benefits span throughout all the functional units of an enterprise due to data sharing. The other functional units of the enterprise realise the benefits mainly due to synergy among various AMTs and simplified data flows among the functional units. Data flow diagram (DFD) is an ideal tool to depict the scale of interaction of advanced technologies among various functional units in an enterprise. By developing a DFD for a company, the decision-maker can gain a comprehensive understanding of the integration aspect of a company. Once the data flow for the whole company is identified and simplified, the necessity to integrate the technologies becomes clearly recognised. The gained knowledge on data flow among various functional units and technologies will then guide the decision-makers to effectively utilise the full potential of the integrated technologies. Integrated manufacturing is particularly important to small and medium sized enterprises. Therefore, a quantitative method should be provided to measure the effect of integration and transform the qualitative information into quantitative data. Then a significant impact on all aspects of the operation of a company with regards to the integration can be obtained. A methodology to evaluate the functional inter-relationships in an enterprise is provided in this chapter. These evaluated functional inter-relationships can be utilised to comprehensively analyse CIM components and the impact on the company in a holistic manner.

2.4 INFORMATION ARCHITECTURE AND PLANNING A CIM SYSTEM

Computer Integrated Manufacturing is a concept which encompasses all aspects of manufacturing related activities. For a successful application the concept requires the most logical and efficient flow of information throughout the company encompassing all activities and integrating the diverse activities including sales, purchasing, design and development, production and quality control, manufacturing, and product delivery. Hypothetically when a customer orders an item, the CIM factory or the "factory of the future" should be able to deliver the product without any human intervention. How far one has progressed or moved towards achieving the futuristic factory is a reasonable question in the minds of the researchers and most of the entrepreneurs. Integrated manufacturing makes good business sense. Making it happen depends upon two important factors, people and computer technologies. The people, whose function is to implement integrated manufacturing, need to understand the precise relationship between integrated manufacturing, AMTs and CIM and be able to exploit it.

Planning an appropriate architecture for CIM plays an important role in integrating the components of CIM in an effective manner to achieve inter operating and working relationship among the integrated network. Architectures can be viewed as a set of instructions that illustrate physical and logical arrangements of the components of the system. It is a formal arrangement that explains functional capabilities, activities, operating boundaries, specifications and physical attributes of the components. Architectures are normally depicted by diagrams to easily understand integration, components and information flow. Any system planning should consider all the components of CIM for better and efficient integrated network. When the planning is carried out in isolation for individual components it is usually found that they do not provide the efficiency required for the entire system and only satisfied the efficiency of that individual component.

Commitment by the management is critical to successful implementation of any innovative technologies. The management approach to integrated systems development can be structured as follows:

1. Analysing the need for integration, understanding the problems associated with it, and defining the requirements of the integration. This should encompass identifying the business functions need to be integrated, identifying integration of an integrated management and information database with other systems, assessing the impact of the investment on the subsystems and on the functional units of the company, identifying the costs related to the procedure and analysing the financial return of the project. In addition, the requirement of the integration should consist of analysing the functions of each subsystem and the data flow among various subsystems in different functional units and developing the relations between the pieces of information and identifying the system performance;
2. Making a preliminary design, formulating the solutions and developing a detailed design for integration. The preliminary design stage should include prioritisation of the needs, tactical planning of the design activities, formulation of the technical approaches and the alternatives, evaluation of the approaches based on technical feasibility, financial feasibility, the

compatibility with the existing facilities and the capability to meet the desired needs within the time frame, and detailing the specifications of the selected priority areas;

3. Building a solution and verifying the results. This should include testing the subsystems and the interfaces, analysing the performance against the original needs and design specifications, and developing logical implementation strategies utilising the current system definitions;
4. Implementing the system and making the user accept it. This should include information secession for the employees, training programs to familiarise with the new integrated system and building the confidence in the system; and
5. Providing maintenance and support. This should be an ongoing process and the facilities have to made to cater for this requirement by having sufficient technical capability in-house or out-sourced, but able to provide support at short notice.

2.5 COMMUNICATION NETWORKS AND PROTOCOLS FOR CIM

Communication networks and protocols are the enabling tools, which allow integration among subsystems of CIM and the functions of the enterprise, for the achievement of the objective. A protocol is a formal set of conventions, which defines the aspect of the communication processes. It defines the format and contents of the message to be exchanged, and procedures to initiate, exchange and terminate data transfer. Communication networks facilitate information sharing among the components of CIM for efficient, fast and reliable interaction. The International Standards Organisation (ISO) developed the Open System Interconnection (OSI) reference model as a general network standard to assist information exchange between distributed systems. The OSI model consists of seven functions (or layers). These functions and protocols are logically grouped in the OSI model to establish communication between different vendor products, thus allowing an open system interconnection approach for applications.

Out of the OSI model, Manufacturing Automation Protocol (MAP) and Technical and Office Protocol (TOP) were developed. MAP was initiated by General Motors Corporation in 1980 as a Local Area Network (LAN) standard to interconnect various computers and control systems of the factory equipment. Similarly TOP was developed by Boeing Company in 1985.

MAP is based on the seven-layer OSI model and uses a token bus, broad band communications network as its physical link. The decision to use the OSI model as a basis to create MAP by General Motors has provided enthusiastic support from computer manufacturers for MAP and this support has enabled MAP to become an accepted industry communication standard for integration.

The intention of the TOP is to be a companion to MAP and assist data communication in office environment and inter-link factory equipment, which uses MAP. Similar to MAP, TOP is also based on the seven-layer OSI model, but it uses baseband as its network specification with a provision for broadband, where an existing broadband currently exists. Although MAP and TOP uses the seven layer OSI model, the first and last layers have some significant differences for both protocols.

2.6 ENTERPRISE MODELLING

Enterprise integration involves data exchange among the various subsystems of CIM in an efficient and effective manner to provide benefits to the enterprise holistically through shared or interconnected databases. Further, an integrated enterprise facilitates co-ordination, communication and co-operation between various business processes of the enterprise including product design functions, manufacturing functions, administrative and support functions and activities related to the external environment and business functions. The main benefit of the integration via computer technology is the right information to the right subsystem of CIM delivered at the right time regardless of the physical boundaries and location of the subsystems. The integration of CIM components are achieved by integrating physical and logical systems with the help of computer communication networks and a variety of network protocols. An integrated enterprise should have the following features:

1. It must logically link different functional areas of the enterprise by common communication protocols;
2. It must have the capability to transmit information vertically from one level to another and horizontally from one function to another without additional work or processing;
3. It must be able to handle the flow of information; and
4. It must have a single data source or interconnected databases and online/real-time updating facility.

A variety of reference models or architectures have been developed over the years to represent components of CIM system and to facilitate the enterprise integration. Some of the widely reported reference models for designing CIM systems are:

1. CALS (Computer Aided Acquisition and Logistics Support), ICAM (Integrated Computer Aided Manufacturing), CAM-I, NBS (National Bureau of Standards) and PERA (Purdue Enterprise Reference Architecture) models developed in the USA;
2. CIM-OSA (CIM-Open Systems Architecture), MMCS (Manufacturing Management Control System), PAC (Production Activity Control), IMPACS and the knowledge-based factory model developed by the ESPRIT program; and
3. GRAI (Groupe de recherche en automatisation integrée) model and GERAM (Generic Enterprise Reference Architecture and Methodology) developed in France.

Many modelling formalisms have been developed within the framework of the above reference models. They include:

1. IDEF (ICAM Definition methods such as Function model $IDEF_0$, Information model $IEDF_1$, and Dynamics model $IDEF_2$) and DFD formalism developed in the USA; and

2. MERISE, GRAI grid, and GRAI net formalism, and GIM (GRAI Integrated Methodology) formalisms developed in France.

Interactions against each functional unit and the technologies can be represented by these modelling formalisms. Out of these modelling formalisms, the DFD method is selected for understanding and evaluating the functional inter-relationships. This particular method is often more easily understood than the other models, in terms of providing both the concept and the configuration.

2.7 FUNCTIONAL INTER-RELATIONSHIP IDENTIFICATION

The success of CIM depends on the ability to link different technologies together in a most effective and profitable method. Enterprise integration contributes significantly in removing organisational barriers among functional units in an organisation. Failing to understand the importance of the information technology as the driving force for the integration is one of the fundamental difficulties in not realising the expected level of benefit with respect to the AMTs. When evaluating interrelated investment projects synergy among the units plays an important role. The complementary technologies have the potential to improve the synergy benefits, while the substitute technologies may not increase the total benefits and may decrease the synergy effects (see Figure 2.2).

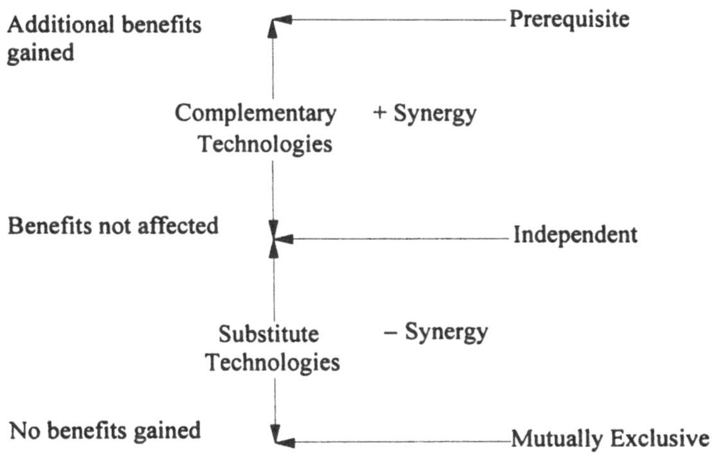

Figure 2.2 Degrees of dependence between interconnected projects [1]

[1] Adopted and improved from Reeve and Sullivan (1990)

Synergy among technologies does not depend on the amount of capital invested, but they rather depend on the ability to share information among the technologies and the ability to impart benefits on other functional units. It is important, in order to satisfy different standards, that a model or methodology developed should be appropriate for different types of industries and businesses. In addition, it should have the capacity to assess the synergy effect of various functional units in an enterprise and of the technologies in the company as a whole. Hence, a generic model in the form of a Functional Inter-Relationship Matrix (FIRM) is proposed. The important difference between the FIRM is that it comprehensively considers the effect of AMTs in various functional units throughout the enterprise.

A properly conceived and clearly explained inter-functional model will provide a consistent mechanism for evaluating possible relationships among various entities and a knowledge in enterprise integration. Thus, in the planning stages of the interrelationship identification, the following has to be verified for every subsystem or island of automation:

- The type of information or data inputs required for a subsystem and the origin of the information; and
- The type of information or data output generated by that subsystem and the need for these data in other subsystems as necessary inputs.

Identifying the communication between the designated subsystems in an enterprise will help in establishing the business and organisational functions of the subsystems and aid in developing the relationships among the subsystems in various functional units. When a technology T_i is implemented in a functional unit FU_j, there will be interaction of that technology on other functional units. Technology T_i impinges on the functional unit FU_j, and at the same time it has an impact on other functional units. The relative effect of the technology T_i on other functional units varies depending on the nature of the functional unit and the technology. The interaction of the technology on other functional units can be described by a diagram, as in Figure 2.3.

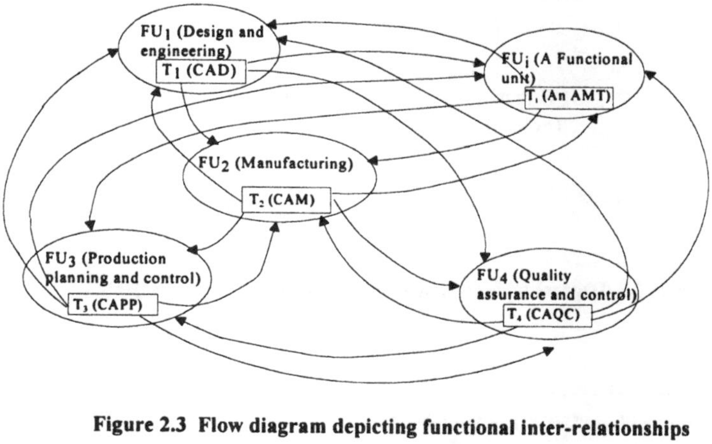

Figure 2.3 Flow diagram depicting functional inter-relationships

To capture the above information and to simplify the process of identification, a tabular matrix with enabling technologies has been proposed. Inter-relationships among various functional units have to be identified in this matrix, by the decision-maker for the investment.

A typical FIRM among technology and functional units is shown in Table 2.1. For example, if a CAD system is implemented in the Design and Engineering unit (D & E) considering it as the prime functional unit, then the relative effect by CAD on D & E is assumed to be 1.0 or 100%. Implementation of CAD in D & E unit would result in a reduction in work related to manufacturing processes in the manufacturing functional unit, due to the flexibility to change and enhanced quality products. Thus, giving a relative effect by CAD on the manufacturing functional unit of about 30%. Similar effects can be estimated by a decision-maker on other functional units.

In this example, the technology of "Office automation" is considered to be implemented in the administration department as the prime functional unit. The manufacturing department is allocated with CAM, GT, CM and FMS as the primary technologies. CAPP is considered to be installed in production planning and control division. Similar identification can be made for the other technologies in the table. The functional inter-relationship identification will reinforce the realisation to the decision-maker that the acquisition of AMTs is not a decision for the functional units concerned, but it is a corporate issue for the company to address. On the other hand, in CIM systems certain elements must be in place for other elements to be effective and the elements that cannot be justified on their own often must be justified on the basis of their contribution to other components of the CIM system. This necessity of having certain elements in CIM systems can be fully realised and justified by the facility of functional inter-relationship identification.

Table 2.1 Matrix for functional inter-relationships

	Functional Units							
Technologies	Admin	D & E	Manf	PP & C	Q A	W & S	Asly	R & D
Office Automation	1.00	0.01	0.05	0.01	0.01	0.20	0.00	0.20
CAD	0.05	1.00	0.30	0.20	0.02	0.06	0.03	0.40
CAM	0.01	0.10	1.00	0.40	0.40	0.01	0.20	0.20
CAPP	0.01	0.01	0.03	1.00	0.20	0.08	0.20	0.20
CAQC	0.01	0.01	0.05	0.07	1.00	0.01	0.01	0.01
AS/RS	0.05	0.02	0.02	0.50	0.10	1.00	0.01	0.01
Robotics	0.00	0.04	0.60	0.05	0.10	0.02	1.00	0.01
GT	0.00	0.01	1.00	0.20	0.10	0.20	0.30	0.20
CM	0.06	0.05	1.00	0.50	0.40	0.08	0.10	0.10
FMS	0.07	0.20	1.00	0.60	0.50	0.20	0.30	0.20

2.8 MEASUREMENT OF SYNERGY

The subjective values on this FIRM can be objectively evaluated by constructing a detailed DFD associated with the activities among AMTs and the functional units. The inter-relationships are dependent on technology, the extent of integration, and the data sharing facility. In the course of identifying the relationships, the decision-maker will learn more about the relationships and limitations. Once the functional inter-relationships are established, it will provide a basis to incorporate the synergy into the justification procedure. Activities are the cost drivers and value generators in manufacturing. Hence, before carrying out a detailed cost and benefit analysis, each activity of the CIM and/or AMTs has to be evaluated to see whether it is a value-adding process. Otherwise, further analysis has to be carried out to see if the process can be eliminated, or simplified. Activities can be made explicit with the DFDs. Once these diagrams have been created, the task of estimating the costs or benefits becomes more systematic as they can be related to the inputs, outputs, and mechanisms which facilitate the activities.

Evaluating a complex integrated enterprise requires an understanding of each component of the integration, and simplifications of the evaluation by appropriate partitioning of the problem. A DFD supports this task by considering activities associated with a single system in an integrated network of DFDs. Later, these components of DFDs can be integrated to gain a comprehensive DFD for the entire enterprise. After gaining the initial understanding of the technologies or the systems invested in an enterprise and then by considering the existing or proposed functional units of that enterprise, the DFD can be constructed for a system by answering following two questions:

1. What type of activities are generated from a system and where can these outputs be connected?
2. What type of activities can be connected to a system and from where do these inputs arise?

As an illustration, consider the technology CAD invested in D & E as the prime functional unit. The features of CAD are:

- CAD systems facilitate preparation of efficient design for manufacturing;
- CAD systems enable designers to optimise and improve their concepts on new products or products being developed, by analysing numerous alternative configurations on the basis of many manufacturing priorities;
- CAD systems can be expanded to generate control procedures for CNC/CAM, to produce process planning for CAPP, etc.;
- CAD systems will have far greater cost impact on functional units such as marketing and manufacturing than they have in the drawing office, since CAD can be used for tasks of producing parts lists, quotations for estimating and generating machine codes for the manufacturing process other than producing drawings as a basic function;
- CAD systems enable designers to produce more accurate drawings and parts list and standardised components. This feature can have significant effects on production costs;

- CAD systems by interacting with the engineering database, can reduce product design and development time. Hence, they can be very helpful for the research and development unit of the company as well; and
- CAD systems, which are available today, provide the facility of having communication with other systems and have an interface to import/export data with other commonly used software products.

Considering the above information, a typical DFD for D & E functional unit which is similar to Figure 2.4 (page 30) can be drawn. The completion of a DFD will facilitate a comprehensive understanding of the activities carried out in the enterprise. It requires a thorough understanding of the enterprise to make it fully effective. Once the DFD is drawn, the functional inter-relationship ratios can be computed by utilising the following steps:

Step 1: For the prime functional unit estimate the following attributes:

- Reduction in time (RT_p);
- Reduction in labour (RL_p); and
- Reduction in indirect costs (RI_p).

Step 2: For every activity denoted in the DFD, estimate all or some of the following attributes:

- Reduction in time obtainable for that activity by utilising the technology i (RT_{ai});
- Reduction in labour obtainable for that activity by utilising the technology i (RL_{ai}); and
- Reduction in indirect costs obtainable for that activity by utilising the technology i (RI_{ai}).

Step 3: For every functional unit j, the functional inter-relationship ratio (aij) associated with the technology i can be computed as follows:

$$a_{ij} = \sum_{j=1}^{n} \frac{RT_{ai}}{RT_p} + \sum_{j=1}^{n} \frac{RL_{ai}}{RL_p} + \sum_{j=1}^{n} \frac{RI_{ai}}{RI_p} \qquad (2.1)$$

Functional inter-relationships may not be constant throughout the useful life of an AMT. Accordingly, even if a value function exists to identify functional inter-relationships, it may change or evolve over time, due to the increased understanding through the learning curve effect. Since the benefits due to integration among various functional units and technologies can be expressed in many terms including that of money, hours, quality, space and weight, the decision-maker should consider these various units of measures before performing the computations and assigning values to the inter-relationship matrix.

If the necessity arises for further improvement, Step 1 can be expanded to cover many relevant attributes instead of limiting with three important attributes. However, in a business decision-making environment, many of the attributes can be computed in terms of monetary units and time. Once the DFD for each technology considered in the enterprise is drawn, the values along the rows of the Table 2.1 (page 27) can be assigned. Then, if necessary, by cascading individual DFD for each technology the complete and comprehensive DFD for the company can be derived.

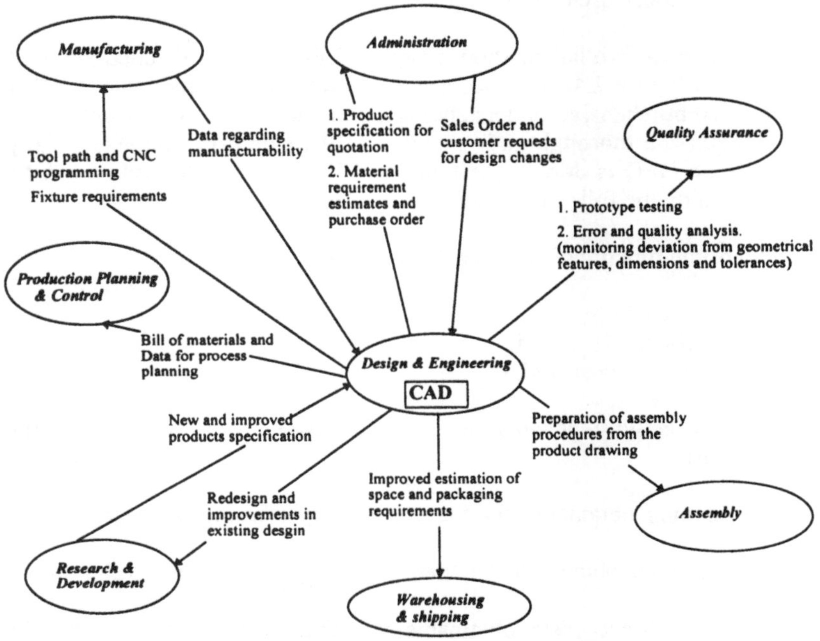

Figure 2.4 A typical DFD for CAD in D & E functional unit.

Synergy for a technology T_i is defined as the sum of functional inter-relationship ratios for that particular technology among all functional units FU_j. Hence, the value of synergy for a technology T_i (θ_i) can be written as:

$$\theta_i = \sum_{j=1}^{n} a_{ij} \qquad (2.2)$$

Effective enterprise integration depends on the ability to identify and to utilise the functional inter-relationships among various units. Consideration of functional inter-relationships among various AMTs considered for investment in an enterprise will provide a mechanism to incorporate the inherent synergy between AMTs and functional units of the enterprise in a justification procedure. The methodology described in this chapter provides a mechanism to objectively evaluate the subjective values of functional inter-relationships with the assistance of DFDs. The derived functional inter-relationship ratios are applied in Chapter five to evaluate the cost-impact and to generate the optimal resource allocation.

2.9 CONCLUDING REMARKS

In this chapter we briefly reviewed some of the advanced manufacturing technologies and enterprise modelling and realised the concept of functional inter-relationships. The functional inter-relationships are relative to the timeframe the estimation is made, since utilisation of cross-functional benefits may not be stationary throughout the useful life of the technology. Accordingly, even if a value function exists to identify functional inter-relationships, it may change or evolve over a time, due to the increased understanding with respect to the learning curve. However, in the course of identifying the relationships, the decision-maker will learn more about the relationships and the limitations faced by the logical and physical configurations of the enterprise. When this limitation is identified, possible data links have to be established with appropriate interconnection standards and software. Realisation of an effective integration by means of efficient communication links among these AMTs will generate significant benefits across many traditional functional units and eliminate organisational barriers among the functional units in an enterprise.

2.10 REFERENCES

1. Appleton, D.S., 1985, *Building a CIM program, A Program Guide for CIM Implementation*, edited by, Savage, C.M, The computer and automated systems association of SME, pp. 5–10.
2. Goetsch, D.L., 1988, *Fundamentals of CIM Technology: automation in design drafting and manufacturing*, Delmer Publishers.
3. Hunt, D.V., 1987, Dictionary of Advanced Manufacturing Technology, (New York: Elsevier Science).
4. Reeve, J.M. and Sullivan, W.G., 1990, "A synthesis of methods for evaluating interrelated investment projects", *Selection and Evaluation of Advanced Manufacturing Technologies*, edited by, Liberatore, J.M, (Berlin: Springer-Verlag), pp. 112–138.
5. Tidd, J., 1991, *Flexible Manufacturing Technologies and International Competitiveness*, (London: Pinter Publishers).
6. Voss C.A., 1988, "Success and failure in advanced manufacturing technology", *International Journal of Technology Management*, Vol 3 No 3, pp. 225–297.

CHAPTER THREE

Developments in Justification of CIM

3.1 OVERVIEW

How do we justify a complex process, when financial consideration plays a dominant role in the determination of the corporate strategy? Even though profitability versus market share, its desire to be a technology leader, internal capabilities and external constraints such as uncertainty in economic, social and political situation influencing the adoption of any investments and particularly AMTs, managers still resort to financial benefits for investment justification. Nevertheless, the economic viability of a project and running a profitable business is the prime concern of any business organisation. Therefore, we should identify a method which brings a positive approach towards investment in AMTs, while addressing the concerns of decision-makers.

Engineering provides many opportunities for investment. However, it is imperative any engineering investment be thoroughly analysed before proceeding to the implementation stages. Benefits and costs of many investments can be quantified in terms of tangible values. Nevertheless, the advanced technologies provide benefits which are intangible and difficult to quantify. In addition, integrated AMTs in the form of CIM will have significant company-wide effects on the organisational culture and the operational requirements. Hence, a comprehensive study has to be undertaken to decide whether CIM has to be adopted without conflicting with the identified corporate strategies and whether it fulfils the criteria of the assessment for an investment. This demands a detailed evaluation of the value of simplified and increased company-wide information flow achieved by CIM. These requirements cannot be met by existing traditional approaches. Hence, a need exists for innovative approaches in justifying advanced technologies.

To identify the shortcomings of the existing approaches, it is essential that we understand the concepts of the traditional engineering methods. Therefore, the primary objective behind this chapter is devoted towards briefly reviewing the available engineering economic tools and exploring the historical developments in justification of CIM. In addition, multicriteria methods such as weighted scoring and analytic hierarchy process are explained.

3.2 ENGINEERING ECONOMIC TOOLS AND INDICES

Engineering economic tools provide a foundation to analyse the economic aspects of engineering. It involves systematic evaluation of costs and benefits of any proposed investment by use of economic indices. Traditional engineering economic analysis includes methods such as PayBack (PB), Return On Investment (ROI), and Discounted Cash Flow (DCF) methods to compute the necessary indices. The DCF method comprises Net Present Value (NPV) and Internal Rate of Return (IRR). The following section gives a short description of the economic methods presently used in industrial applications. However, the reader who is willing to have a detailed knowledge in engineering economics is advised to read books specifically dealing with engineering economics.

3.2.1 Payback Method

Payback measures the time period to recover the initial investment. The PB method, due to the simplicity of the analysis is a widely used financial appraisal mechanism either in the form of a discounted PB or as a non-discounted (simple) PB. The simple PB method calculates the number of years required to reach a positive cash flow that equals the total investment. The simple PB method is widely used in industry, since the method is easy to use and understand in comparison to the preferred DCF techniques. The disadvantage of the simple PB method is that it does not account for the monetary value of time, whereas the discounted PB method considers the monetary value of time. Many researchers have expressed criticism against the simple PB method. The use of the simple PB method for the evaluation of new technologies may lead to the rejection of investment in AMTs. Some researchers argue, to achieve integrated manufacture that promises a path towards minimum overall product cost, better response and improved quality, the PB method should be removed from the evaluation techniques. However, many others propose the use of PB as a secondary method in conjunction with DCF as a primary capital budgeting technique, since PB helps understand the time required to recover the capital investment.

3.2.1.1 Simple Payback

The simple PB method computes the time to recover the committed investment on the project through the net positive cash flows within the study period (N). Ignoring the time value of money, the net positive cash flow is computed by subtracting net expenditure of the j th period (E_j) from the net revenue of the j th period (R_j).

The simple PB period m ($m \leq N$) can be calculated from the following expression:

$$\sum_{j=1}^{m}(R_j - E_j) - I \geq 0 \tag{3.1}$$

Where

E_j = Net expenditures including investments for the j th year
I = Initial capital investment
N = The number of units of study periods
R_j = Net revenues or savings for the j th year.

However, when J = N, the salvage value of the project is also considered in the computation of the PB period.

Example 3.1
An investment of \$20,000 is made in a company, which incur annual expenditure of \$3,000 in the form of maintenance, overheads and operation costs. The study period of the project is six years. The project can produce an annual revenue of \$7,500 for first three years and \$7,000 for remaining three years. The salvage value of the investment at the end of the sixth year is \$4,000. By utilising a table as shown in Table 3.1, the simple PB can be easily computed. (Please note that inflation, depreciation of assets and income-tax are not considered in all these examples.)

Table 3.1 Computation of simple PB period

Year (J)	Rj	Ej	Net cash flow	Cumulative cash flow
0		20,000	−20,000	−20,000
1	7,500	3,000	4,500	−15,500
2	7,500	3,000	4,500	−11,000
3	7,500	3,000	4,500	−6,500
4	7,000	3,000	4,000	−2,500
5	7,000	3,000	4,000	1,500
6	11,000	3,000	8,000	9,500

From the cumulative cash flows, we can say that the PB period should lie between four and five years. Then by prorating the cumulative amount of these adjacent values to reach zero, a partial payback period can be found. Partial payback period = 2,500/(2,500 + 1,500) = 0.63. Then the PB period for the investment is 4.6 years.

3.2.1.2 Discounted Payback

Discounted PB method is used to account for the time value of money. In this method the monetary value of the future cash flows are converted to the present value by a minimum attractive rate of return (MARR) expected by the company on the investment. Utilising the notations of Equation 3.1, the discounted PB period m ($m \leq N$) can be calculated from the following expression:

$$\sum_{j=1}^{m}(R_j - E_j)/(1+i)^j - I \geq 0 \tag{3.2}$$

Where
  i = Effective interest rate per interest period (MARR).

Example 3.2
The minimum attractive rate of return considered by the company for Example 3.1 is 10%. By utilising a table as shown in Table 3.2, the discounted PB can be easily computed.

Table 3.2 Computation of discounted PB period

Year (J)	Rj	Ej	Net cash flow	Present worth of Net cash flow at 10 %	Cumulative cash flow of present worth
0		20,000	−20,000	−20,000.00	−20,000.00
1	7,500	3,000	4,500	4,090.91	−15,909.09
2	7,500	3,000	4,500	3,719.01	−12,190.08
3	7,500	3,000	4,500	3,380.92	−8,809.17
4	7,000	3,000	4,000	2,732.05	−6,077.11
5	7,000	3,000	4,000	2,483.69	−3,593.43
6	11,000	3,000	8,000	4,515.79	922.36

From the cumulative cash flows, we can say that the PB period should lie between five and six years. Then by prorating the cumulative amount of these adjacent values to reach 0, a partial payback period can be found. Partial payback period = 3,593/(3,593 + 922) = 0.80. Then the PB period for the investment is 5.8 years.

3.2.2 Discounted Cash Flow

As explained in the discounted PB method, discounted cash flow methods measure cash flow with the consideration of the monetary value of time, since the money borrowed today is not equal to the amount which has to be repaid in future due to the interest rate on the capital. Similarly, money which will be available in future will have less value in today's terms. In addition, inflation changes the value of the money in time. To accommodate monetary value of time DCF methods can be utilised in the financial analysis. DCF methods rank second in the usage as appraisal mechanisms in industries and include equivalent worth methods and IRR. The equivalent worth can be computed by present, future or annual worth methods. Among the DCF methods, the IRR method is widely used in industry.

The distinction of the IRR method is it computes the rate of return from the investment, whereas the other methods convert the cash flows resulting from the investment to the equivalent worth at some point in time with the arbitrarily chosen MARR of the company.

3.2.2.1 Internal rate of return

The IRR method involves calculation of an interest rate that equates the equivalent worth of cash inflows in the form of receipts or savings by the investment in consideration to the equivalent worth of cash outflows in the form of expenditures including investments. IRR helps to benchmark the return of the investment against an MARR, which can be regarded as a minimum standard for a company. The advantage of the IRR method is that it takes into account the effects of timing on cash inflows and outflows related to that investment throughout the study period considered.

According to the above notations

$$\sum_{j=0}^{N} R_j/(1+i^*)^j = \sum_{j=0}^{N} E_j/(1+i^*)^j \tag{3.3}$$

When Equation 3.3 satisfies, the i* is the IRR of the investment and when i* is greater than or equal to the company's MARR, the investment is financially viable.

Example 3.3

In this example the IRR for investment discussed in Example 3.1 is computed. The IRR can be calculated by a trial and error method by assuming reasonable interest rates for i* and interpolating or by using iterative techniques. In this exercise Microsoft Excel software is used to compute the IRR. In this example, IRR = 11.5%. This is slightly higher than the MARR of 10.0% considered by the company.

3.2.2.2 Net present value method

Net present value is the other widely used DCF method used for financial evaluation. NPV is called a net present worth method as well. It returns the net present value of the investment based on a series of periodic cash flows and a discounted rate. The NPV method is closely related to the IRR method. Generally, the discount rate of NPV to compute the equivalent worth of the future cash flows is an arbitrary interest rate called MARR. NPV is derived by discounting future net cash flows by MARR, summing them over the life of the investment, and deducting the initial investment related to the proposal. When the NPV yields a positive value the proposal is considered to be economically satisfactory. In other words IRR is the interest rate for which NPV becomes zero.

$$NPV = \sum_{j=0}^{N} \frac{(Rj - E_j)}{(1+MARR)^j} \tag{3.4}$$

Example 3.4
In this example the NPV for investment discussed in Example 3.2 is computed. The last column of Table 3.2 is useful in getting the NPV for the investment. Each row of that column gives the NPV for the investment at the end of each financial term. The last row gives the NPV for the investment at the end of the project. In this example, NPV for the project = $922.36.

The IRR and NPV methods place stronger emphasis on short-term benefits by discounting future cash flows heavily. Hence, the application of these methods for the evaluation of an AMT without recognising the qualitative benefits of the AMT will result in rejection of the AMT investment. In addition, DCF techniques as the only financial analysis tools are inadequate for evaluating investments, and an over emphasis on it leads to under investment in manufacturing facilities.

3.3 EXISTING APPROACHES FOR AMT/CIM JUSTIFICATION

Justification of CIM is an integral part of the CIM planning process. Installing CIM or any subsystems of CIM such as CAD, CAM, CAPP, CAQC and others in a CIM environment is one of the most lengthy, expensive, and complex tasks an organisation can undertake. In addition, new technologies are considered by decision-makers in manufacturing industry to achieve diverse goals of the organisation and to satisfy various customer requirements. The customer requirements may include criteria of shorter delivery time, competitive pricing, improved product quality and reliability, diversity of products to meet the product life cycle, and improved product innovation. Hence, the justification of CIM has to accommodate these multicriteria requirements for a proper evaluation.

Justification approaches have been categorised in a variety of ways. Boddy and Buchanan (1986) elaborated that justification falls between the extremes of the opportunistic and strategic approach, depending on the style of management. Sullivan (1986) distinguished between the tactical and strategic approach and stressed that the justification which involves the increased use of automation should rely on a strategic approach. Boaden and Dale (1988) had suggested that the justification of CIM requires both concept and equipment justification. Concept justification verifies whether CIM fits into the corporate strategy, whereas the equipment justification is directly concerned with cost justification of the individual elements of CIM. Mohanty (1993) argued that classifying the literature according to the method of analysis, such as qualitative methods, semi-quantitative methods and quantitative methods could lead to a better understanding in terms of issues that require subjective or objective consideration.

Many approaches and methodologies have been proposed by researchers and consultants to solve the problems in justification. Meredith and Suresh (1986) categorised the available approaches for justification in investment in AMTs as follows:

1. Economic justification approaches;
2. Analytic justification approaches; and
3. Strategic justification approaches.

Their aim was to highlight that the justifications for AMT involve strategic factors, which are often difficult to quantify, rather than cost reduction factors and recommended the combination of all three approaches to analyse the impact of decision-making. Later, Krinsky and Miltenburg (1991) stated that in order for an organisation to be effectively managed, it is necessary to perform both strategic and economic analysis simultaneously. Based on these analyses an organisation can positively identify whether or not it needs to invest in AMT.

McKay and Galbraith (1989) classified the justification approaches with regards to the use of accounting techniques, into three schools of thought. The first group argues that current accounting systems are adequate and that justification can be achieved through detailed analysis with the quantification of intangible benefits and their subsequent incorporation within the overall justification process. The second group considers the strategic issues as the prime function in the justification process for the advanced technologies. In addition this group disapproves of the use of conventional accounting techniques, and emphasises long-term considerations to gain benefits. The third group considers the AMT investment to be significantly more complex in its basic nature than straightforward capital investment calculations, and as such requires a more complex and rigorous method of numerical analysis. These methods recognise the wider benefits of AMT and attempts to incorporate them into the justification process. The present day approaches can be categorised as shown in Figure 3.1.

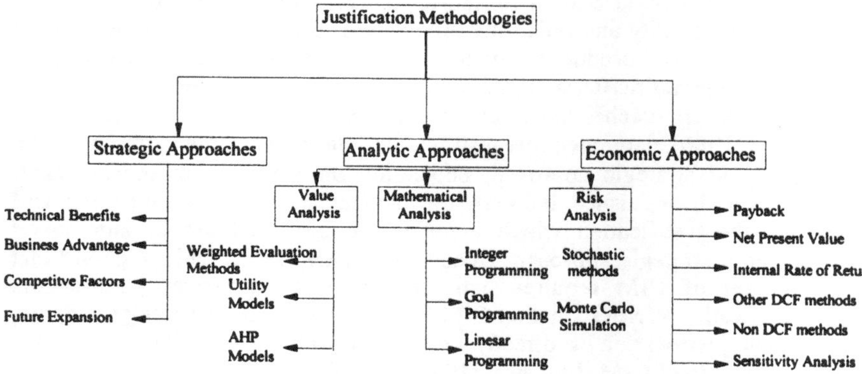

Figure 3.1 Available justification approaches

3.3.1 Economic Justification Approaches

Economics of FMS justification has been pioneered by Hutchinson (1976) and he has evaluated the economic advantage of flexibility by comparing transfer lines and flexible systems through a simulation approach. The economic justification approach analyses the justification issue with simple economic functions such as PB, ROI, IRR, NPV and others. These approaches cannot analyse the non-economic, strategic benefits but have the capability to analyse issues where there is no uncertainty such as in stand-alone equipment justification or replacement justification. The advantage of these economic models lies in their simplicity and their ability to identify profitability of the investment which is the bottom line for any capital investment of an organisation. However, some researchers advise the manufacturing managers to disregard financial measurements and make decisions based entirely on an analytical approach.

3.3.2 Analytic Justification Approaches

Analytic justification approaches are appropriate tools for analysing systems which have economic and non-economic benefits. These approaches offer a realistic solution to knowledgeable decision-makers, when sufficient information is available for multiple-attribute justification, while requiring more time from managers for the analysis. These analytic approaches help the decision-makers prioritise the attributes which are desirable to the company. These methods can be utilised to analyse the option of linking advanced technologies with the existing ones. The analytical approaches can be further subdivided into three categories, such as value analysis, mathematical analysis and risk analysis.

3.3.2.1 Value analysis

This analysis consists of finding the value of the investment using weighted scoring models, utility models and analytic hierarchy process (AHP) models. The weighted scoring model is basically performed by assigning a weight to the set of parameters, which are already identified. Then the scores are summed up to get the total score of the project. The total score is utilised to select the best project among alternatives which are under comparison.

3.3.2.2 Mathematical analysis

Mathematical programming models provide a basis to determine the best course of action to a decision-making problem. It is applicable to a decision-making problem when the objectives and the constraints of the problem can be expressed mathematically as functions and decision parameters. The mathematical analysis includes linear, integer and goal programming algorithms to solve multi-attribute problems. This approach is mainly considered as an optimisation procedure using such programming models and is included as a component in the justification procedure.

3.3.2.3 Risk analysis

Over the years, risk analysis has become an important evaluation tool to assess investments. This approach popularised by D. B. Hertz for capital investments in AMT relies on simulation of probabilistic factors to determine the outcomes in statistical or graphical models. Risk analysis is a good analytical tool to explore the optimistic and pessimistic estimates of a decision process and simulate the outcome of the decision. In traditional approaches, risk is generally incorporated into the interest rate to provide an allowance for wrong decisions. Introducing high interest rates has been criticised by almost all researchers. Since CIM technologies are very complex and impact on many areas of the organisation, risk analysis should be incorporated in an evaluation process. Risk analysis is a complex analytical process and requires a huge amount of data for the complete analysis. However, it offers a facility to estimate the possibility of achieving various measures including cost, profits, market share, yield, and capacity, when a deficiency in precise knowledge exists in a decision-making process.

3.3.3 Strategic Justification Approaches

Strategic justification approaches are less technical in nature compared with other approaches but they highlight qualitative attributes including business strategies, flexibility in meeting customer demand, and competitive advantage. These approaches identify long-term goals of a company. The success in adopting an AMT increases with greater possibility if the decision to implement an AMT relies on strategic considerations. When the investments in AMT or CIM are the crucial means of sustaining the competitive position of a company, strategic arguments should dominate the investment decisions, and sometimes even forgoing immediate and short-term financial benefits. A strategic approach is necessary for the success of adopting innovative technologies with the participation of all concerned in the implementation and usage of the technology. For an integrated technology such as CIM, the strategic approach will bring a better result if it is used in conjunction with the economic models as a component of a multicriteria method.

3.3.4 Multicriteria Methods

In today's competitive world the presence of multicriteria is a standard feature in almost every aspect of management decision-making. The science of Multiple Criteria Decision-Making (MCDM) evolved as a separate field of study after the Second World War. During and immediately after the War, the management scientists were primarily concerned about solving problems with single predefined criterion. That pattern of research satisfied the need of that day by providing tactical answers for national emergency issues. However, the issues which require multiple dimensions or criteria are all around us in most day-to-day activities.

Multicriteria situations arise when a situation has to simultaneously address multiple goals. Since, implementation of an AMT involves satisfying diverse goals, the investment for an AMT should be justified on the basis of multiple objectives rather than on a single objective such as the maximisation of return on investment or the minimisation of the PB period. Many multicriteria methods have been developed over the last decade to address the justification of CIM or any issue with multiple attributes in consideration. The use of multicriteria methods help decision-makers in solving problems involving multiple objectives by providing a framework and techniques to obtain feasible solutions in accordance with the preference of the decision-maker. However, research striving for an optimal solution within MCDM is still in its infancy.

Various multicriteria methods are available in literature, including graphical techniques, ranking of attributes, scoring models, utility models, fuzzy techniques, outranking methods, analytical hierarchy process, and multi-objective mathematical programming techniques such as goal programming.

The graphical techniques include attributes score card, shaded circles to portray scorecard results, graphs of evaluation ratings for attribute outcomes and polar graphs. Ranking of attributes involves ranking the attributes in the order of decreasing preference. Scoring models involve quantifying the relative importance of the attribute by assigning weights. Various types of scoring models exist including unweighted 0 to 1 factor models, unweighted models and weighted scoring models. For assigning weights, the models such as numerical formula methods including uniform and equal weights, ranks sum weights and rank reciprocal weights, could be utilised. Utility models express the preference of the decision-makers on the combination of the outcomes either by additive or multiplicative function of the individual attribute utility functions. Fuzzy methods involve the use of linguistic variables from the theory of fuzzy sets by converting verbal expressions into numerical ones to evaluate the CIM strategy. Outranking approach consists of developing a set of alternatives for multicriteria, to represent the preferences of the decision-maker. AHP consists of hierarchical representation of the decision problem and makes pairwise comparisons among the factors to derive the best alternative for the problem. Among all these methods the weighted scoring model and the AHP are widely used in solving multicriteria problems. The authors of this book propose those two models for AMT and CIM justification.

3.3.4.1 Weighted scoring technique.

The weighted scoring technique is considered as a conventional and highly useful multicriteria decision-making tool. The technique provides a systematic subjective quantification process to evaluate the alternatives according to the performance measurement scale of the criterion and enables a decision-maker to determine the level of contribution by every criterion towards satisfying each alternative.

To be considered as an MCDM technique, the evaluation techniques should have the following characteristics:

- Identifying the relevant and important system evaluation criteria;
- Determining the importance of each criterion;
- Evaluating the various alternatives with respect to each criterion; and
- Performing appropriate calculation and identifying the preferred alternative for the company.

The weighted scoring technique completely satisfies all the above requirements to be an MCDM technique. The weighted scoring technique consists of initially identifying a list of attributes and alternatives for an evaluation and then assigning weights for the attributes on the scale of 0 to 100. Assigning weights on this arbitrary scale is a difficult task but it provides a mechanism to evaluate the relative value of the alternatives under consideration. Generally, the attribute which is perceived as the most important to meet the predetermined objective is assigned the highest rating and all other attributes are assigned with weights relative to the highest one. If the ranking of attributes is determined by the method of paired comparisons or by any other means, it will assist the decision-maker to assign the relative weights for the other attributes.

Once the relative weights are assigned, the alternatives are rated on their ability to satisfy the performance requirements with respect to each attribute. Then the weights are normalised by dividing the weight of each attribute by the sum of all the weights. Each evaluation is then weighted by multiplying the weight of the attribute by the corresponding rating for the alternative. Finally the overall score of an alternative is determined by summing its weighted evaluation for all the relevant criteria. Then the alternative with the highest score is chosen for the further consideration.

3.3.4.2 Analytic hierarchy process

Analytic Hierarchy Process is a methodology developed by Saaty (1980) to analyse rational and irrational values comprehensively according to the level of importance to the decision-making process. AHP facilitates formulating and simulating the human decision-making mechanism in multicriteria evaluation procedures. In addition, it is an effective mechanism to analyse the strategic concepts of a company by the representation of a complex problem into a disintegrated hierarchical problem. This disintegrated representation of multiple level hierarchy helps the decision-makers identify the problem and deal with it in a clear and concise manner. The complexity of the problem and the thoroughness required to arrive at a viable solution determines the number of levels of hierarchy and the number of elements within each level of the hierarchy. The top level of the hierarchy consists of a single element, which is the main focus of the overall objective. The remaining levels may consist of few elements. Generally, the number of elements in each level will be between three to nine. AHP compares each element in each level with each other element by a pairwise comparison process, with respect to the objective. The pairwise comparison is done through the subjective evaluation of the decision-maker depending on the nature of the investment and the importance of the attributes to a company.

A matrix is constructed by listing the attributes to be compared to the left of the row and to the top of the column. The attributes are compared along each row with the attribute on the column. When an attribute is compared with itself the value on the cell is assigned to one. When an attribute is compared with another attribute, the value is assigned depending on the importance of that attribute to the compared one to meet the objective. If that attribute is more important an integer value is assigned. If that attribute is less important, the reciprocal value is assigned. The reciprocal value is entered in the transpose position of the matrix. For a square matrix with n number of variables, it is necessary to have $n(n - 1)/2$ pairwise comparative judgements for the calculation of the relative importance.

The AHP procedure recommends a 1 to 9 scale to compare the attributes. When attribute A is compared with attribute B, the intensity of the scale may be assigned according to Table 3.3 (page 43).

Once the matrix has been completed, the priority weights for the matrix are computed. In mathematical terms it is called principal eigenvector. The estimate for that vector can be computed in the following four ways (Saaty, 1980):

1. The values in each row are summed together and that summation is normalised by dividing each sum by the total of all the sums, resulting in a summation of the vectors to unity. The first resulting vector is the priority weight of the first attribute, the second value for second attribute and so on;
2. The reciprocal of the sum of the values in each column is computed. Then each reciprocal is divided by the summation of all reciprocal values, resulting in summation of all vectors to unity, thus obtaining normalised values. Then the priority weight is determined as in the first method;
3. Normalisation of the column is carried out by dividing the values in each column by the sum of that column. Then the elements in each resulting row is added and that value is divided by the number of elements in the row, thus achieving the process of averaging over the normalised columns; and
4. The **n** number of elements in a row are multiplied and the $\mathbf{n}^{th}$ root is calculated. Finally the resulting numbers are normalised, to get the priority weight of each attribute.

Table 3.3 Scale of relative importance for AHP matrix

Intensity	Definition	Explanation
1	A and B are equally important	A and B contribute equally to the objective
3	Weak importance of A over B	Experience and judgement slightly favours attribute A over B
5	Essential or strong importance of A over B	Experience and judgement strongly favour attribute A over B
7	Very strong or demonstrated importance of A over B	Attribute A is strongly favoured over B, and the dominance of A has been demonstrated in practice
9	Absolute importance of A over B	The evidence of favouring A over B has the highest possible order of affirmation
2, 4, 6, 8	Intermediate values	When compromise is needed

The accuracy of the result for the priority weight of an element improves on the ascending order of the mechanisms shown above. That is, the fourth method offers a superior technique of estimating the priority weight than the first method. The process of comparing the criteria by pairwise comparison and computing the priority vector for each matrix is carried out starting from the top level of the hierarchy and continuing until the lowest level is reached. Inconsistency is a flow in the human decision-making. The AHP procedure by providing a consistency index for the verification of the consistency ensures consistent and logical evaluations. The consistency of the values is calculated using the principal eigenvalue of the matrix (λ_{max}). The approximate value of the λ_{max} can be obtained as follows:

1. Multiply the matrix of comparison [A] on the right by the priority weight vector matrix [B] on the left. This will result in a new column vector [C];
2. Divide each component on the vector C by the corresponding component in the priority weight vector B. This will result in a new vector [D]; and
3. Divide the sum of the components in the vector [D] by the number of components in the vector [D]. The resultant value is approximately equal to the λ_{max}.

The deviation from the consistency is called a consistency index (CI), and it is represented by $(\lambda_{max} - n)/(n - 1)$. The consistency indices were randomly generated for reciprocal matrices with reciprocals forced to the 1 to 9 scale and of the order 1 to 15 using the sample sizes of 100 and 500 (Saaty, 1980). These generated values are labelled as a random index (RI) and given for a matrix of order 1 to 15. The ratio of CI to the RI for the matrix of the same order is called a consistency ratio. The pairwise comparison is considered satisfactory if the consistency ratio of the comparison falls below or equal to 0.10. The measure of consistency by the consistency index enables the decision-maker to improve the overall consistency of the comparative judgements.

Finally, the weighted evaluation for each alternative is obtained by multiplying the matrix of evaluation ratings of the alternatives with the matrix of priority weights of the attributes, moving upwards through the hierarchy and summing over all the vector values for the alternative. The alternative with the highest weighted evaluation is considered to be fulfilling the objective of the problem with maximum satisfaction and chosen for further consideration. Figure 3.3 illustrates the principle of vector multiplication considering two alternatives A and B, which are taken for the evaluation. The value given next to each attribute represents the priority weights for the attributes.

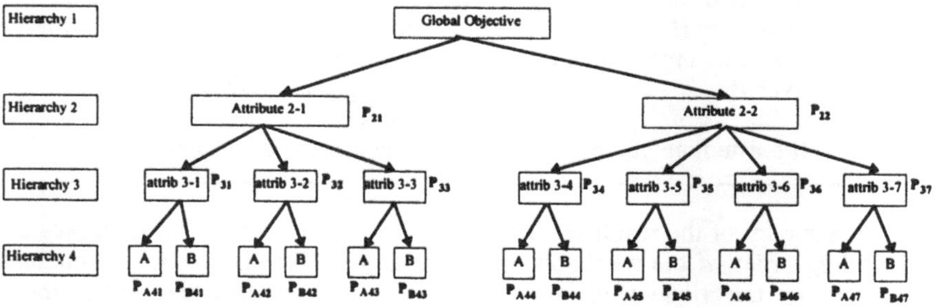

Figure 3.2 Hierarchies of AHP

For A, the summation of the priority weights $= \sum_{i=1}^{3} P_{A4i} * P_{3i} * P_{21} + \sum_{i=4}^{7} P_{A4i} * P_{3i} * P_{22}$

For B, the summation of the priority weights $= \sum_{i=1}^{3} P_{B4i} * P_{3i} * P_{21} + \sum_{i=4}^{7} P_{B4i} * P_{3i} * P_{22}$

3.4 SHORTCOMINGS OF THE TRADITIONAL DECISION-MAKING TECHNIQUES

Many companies establish their investment decisions by traditional discounted cash flow financial justification methods. These traditional methods are mostly suitable to meet profitability criteria rather than long-term strategic goals. Traditional financial methodologies, which are often coupled with high hurdle rates, result in avoiding AMTs in manufacturing enterprises.

In a manufacturing environment any investment has to be justified to satisfy the strategic or financial objectives of a company before being implemented. The tangible benefits and costs are the standard measures used by all companies to justify an investment, but as advanced technologies become more progressive and capital intensive, companies are now forced to look into benefits other than direct cost savings. The benefits offered by the recent developments in AMTs cannot be justified by the traditional financial methodologies. Almost all the articles dealing in the evaluation of AMT/CIM criticise the inability of conventional accounting techniques for the justification of AMT and CIM. This fact has been widely accepted among researchers and consultants in the area of justification. The inability of the conventional appraisal methods can be summarised as follows:

1. They give a high priority to direct labour savings thereby implying that the investment is a threat to jobs;
2. They do not concede the company wide benefits and considerations are only given to the direct benefits of AMT;
3. They do not concede the interdependence of departments and the synergy among various technologies in a company;
4. They assume the impact of investment is limited only to the immediate environment of the equipment;
5. They do not have the ability to evaluate all the potential benefits of an AMT, where most of the benefits are qualitative;
6. They ignore the possible revenue enhancements by AMTs;
7. They assume same market share with the new investments;
8. They expect the market share to remain as it is disregarding competitors ability to invest in AMTs;
9. They are coupled with high hurdle rates and short recovery periods;
10. They consider the capabilities of the equipment are well known and fixed or consider that they decline over a period of time, thus do not consider the positive learning curve effect of AMTs;
11. They fail to adequately quantify and incorporate the broad and pervasive benefits of CIM; and
12. They often blind firms to the effects of status quo complacency.

However, the difficulty rests not on the financial techniques but in the poor application of these techniques to investment proposals (Kaplan, 1986). Consultants and researchers urge manufacturing managers to apply financial analysis in a more appropriate and sensitive manner considering the realities and special attributes of CIM and not avoid them, in CIM justification.

The importance of CIM to the organisations is, CIM brings many tangible and intangible benefits. However, most of the benefits are in the form of intangibles and they are naturally very difficult to quantify. Besides this issue, CIM differs from traditional capital investments in several important ways in addition to the size and complexity of the CIM programs. The variations can be summarised as follows:

1. Benefits gained due to integration of technologies play a vital role in CIM and the benefits of CIM are not limited to the initial positive impact of the equipment on the organisation. As the CIM program proceeds, synergy resulting from the integration of equipment and systems will continue to be identified due to the positive learning effect. Hence, benefits of CIM will increase with time and expand due to integration-synergy;
2. The cost of CIM can decrease with time due to the increased compatibility in hardware and due to the optimisation of system software performance during its life;
3. CIM provides many intangible benefits including improving the effectiveness of communications with suppliers and customers, and the ability to respond rapidly to new opportunities and problems. These benefits help increase the competitive capability of the company; and
4. The effect of CIM in the organisation is far reaching. Effective management of CIM requires new skills and attitudes from people in the organisation. In order to be successful, the management should expect organisational changes at a very early stage in the program and should provide facilities for human resource development.

These considerations necessitate an appropriate justification mechanism before the adoption of CIM in an enterprise. Traditionally, manufacturing machinery was justified on the basis of labour and cost reduction. However, AMTs offer strategic benefits far more than cost reduction alone. It is also important to recognise that there is a possibility the CIM may not give the expected financial benefits on a short-term financial horizon. Justifying strategic benefits and assessing financial implication of the investment in CIM has become a difficult process for the decision-makers of a company. In addition, incorporating synergy among various functional units of a company due to CIM, and strategic benefits into the decision-making process puzzles the decision-makers, since they have been justifying investments purely based on financial returns. These complex issues in the justification of CIM require an innovative methodology to make decisions. The authors of this book have proposed an innovative approach for justifying CIM and AMT and a decision support system, which overcomes above shortcomings, to help manufacturing managers resolve the challenges faced in justifying CIM implementation.

The traditional decision-making techniques that are being used to evaluate a company's investment include PB method, IRR, NPV, cost benefit ratios and risk analysis. These traditional methods are very well known and applied in industry to evaluate investments. However, these traditional approaches are quantitative analysis and do not have the flexibility to recognise many important qualitative factors of any AMT investments. In addition, a complex decision of AMT investment with multicriteria cannot be reduced to a single parameter decision.

While most of the benefits from AMT and CIM investment are intangible or "non financial", over the long-term these benefits bring financial benefits to the company. However, the traditional techniques have no facility in evaluating them. Traditional appraisal methods when applied for AMT evaluations impose severe limitations on AMT, which often result in wrong investment decisions being made and the wrong decision leads to the rejection of the proposed AMT to be implemented. The major shortcomings of the traditional investment evaluation techniques are:

1. All the traditional mechanisms narrowly focus on cost savings and maximising short-term returns instead of benefit appraisal which tends to inhibit investment in AMT;
2. Many of the traditional cost measures do not evaluate the strategic implications of CIM or AMT investment which do not have clearly defined economic benefits; and
3. Many of the justification approaches do not provide facilities to analyse the multicriteria problems, evaluate intangibles and synergy, and incorporate them in the evaluation process in an integrated manner.

Due to these considerations, the traditional investment appraisals have become a barrier to the adoption of new manufacturing technologies. There are many academic articles giving suggestion on the implementation of AMT. These suggestions give varieties of methodologies to manufacturing managers and do not give an approach which guides the managers along the implementation process. The authors of this book believe that the literature should not limit itself only to providing suggestions on how to implement AMT or what to do with an AMT investment, but the literature should provide mechanisms and decision support aids to improve the manufacturing environment by investing in AMTs.

There are a number of methodologies for investment appraisal and software tools such as AutoMan (Weber, 1992), Expert Choice (Radcliff, 1986), and Prefcalc (Radcliff, 1986) for evaluation of AMTs. All these approaches fail to offer a better approach to quantify the intangibles in monetary terms for the decision-maker. Today, a number of researchers are developing decision and management science models to address the strategic framework of manufacturing. Whilst each new methodology is an improvement on preceding approaches, an appropriate answer to quantify intangibles and to address the functional inter-relationships has still not been found.

3.4.1 Dilemma in Making Manufacturing Decisions

The operating environment of manufacturing is changing dramatically. The features of the traditional manufacturing environment such as product stability, domestic markets with few competitors and predictable economic trends have disappeared. The manufacturing environment is now characterised by shorter product life cycles, global and more specialised markets and increased competition from both local and foreign manufacturers. In addition, the introduction of AMTs has considerably changed the cost-mixture of investment analysis. Software and accessories costs plays a major role in the total cost of AMT, thus reducing the significance of traditional cost information such as direct labour and depreciation. Furthermore, the use of traditional financial techniques, which are more suitable to meet short-term profitability criteria, is no longer adequate to be used in a multicriteria decision-making environment and could not incorporate intangible benefits and costs for a complete evaluation of AMTs.

All those above issues pose a dilemma for manufacturing managers to make investment decisions regarding AMTs. Furthermore, absence of adequate analysis and suitable justification methods is a major problem in dealing with investment decisions for CIM and AMT. To overcome this predicament many researchers focused their research activities towards finding suitable mechanisms for justifying investments in AMTs.

Numerous articles have been written regarding CIM and AMT justification. Much of the literature on justification of AMTs are of a qualitative nature and do not provide any clear path towards justification. The literature of AMTs often sites advantages and the qualitative attributes of an AMT implementation to a company. While this information is helpful, the qualitative attributes of these AMTs should be quantified and measured. Due to the wide variance in the reported benefits, each AMT procurement decision should be based on the circumstances relevant to the company alone and most importantly be based on the strategic requirement of the company.

3.5 AN INSIGHT TO AVOID SHORTCOMINGS

How do we avoid shortcomings of the traditional evaluation techniques is a question in our mind now. The shortcomings of the traditional approaches can be avoided, if we incorporate facilitates to quantify intangibles and include them in the evaluation procedure, evaluate and incorporate synergy benefits in justification, and optimise resource allocation based on a strategic concerns of the company in an integrated approach. In addition, our integrated approach should have facilities to operate under an MCDM environment. The first step of solving this complex problem is to identify approaches which can at least solve individual problems. Then we could identify some methodologies to integrate all individual solutions into a single and integrated justification and optimisation approach for CIM.

If the value of intangibles could be assessed, we could quantify and incorporate them into the justification procedure. The Cost-Value technique is one such method to assess the value of the intangibles. The next section deals with value measurement techniques and in particular the Cost-Value technique.

How do we evaluate synergy? In Chapter two we have already identified a technique to do that. An MCDM technique for optimisation is analysed in the next chapter, which we could use to derive an optimal resource allocation for the investment evaluation with multiple issues. Then we could formulate an appropriate justification and optimisation approach.

3.6 VALUE MEASUREMENT TECHNIQUES

Generally, anything capable of being appreciated could be considered to have a value. Value of an object is related to the importance attached to that particular object. Measurement of a value is relative rather than absolute and depends on the perception of the people and the attributes they desire. All the qualitative attributes related to an AMT improve the value of that technology. Naturally, values are intangibles.

Value engineering and value analysis are two formal methodologies considered in the study of value. Value engineering deals with issues involving the products whereas value analysis deals with issues pertaining to management and administrative systems. Value engineering evolved as a technique in finding suitable substitute materials during the Second World War when General Electric Company assigned Mr. Lawrence D. Miles, an electrical engineer, to purchase critical materials for its purchasing department. He discovered that the performance of materials, components, and products could be improved while reducing the cost by understanding and addressing the intended function of the design. Miles named this discipline as value analysis. When the USA navy applied this technique later, it changed the name to value engineering. However, the principles applied to both disciplines are the same.

Many value measurement and value screening techniques are used to measure the impact of the values in a decision-making process. Value measurement techniques are utilised to evaluate the relative measure of importance of the attributes depending on the perception of the decision-maker. These techniques include simple ranking, weighted ranking, successive ratings, pair-comparisons such as regular-pair and scaled-pair comparison, pair-cost-comparison, direct magnitude estimation, category scaling, criteria analysis and scoring models. On the contrary, value screening techniques are utilised to qualitatively control and significantly reduce the large number of parameters under consideration. Value screening techniques includes pareto voting, ranked pareto voting, and Q-sort.

Simple ranking is the most commonly utilised method and it involves ranking the list of items according to the order of importance in a sequence. Successive ratings involves assigning weights using a 0 to 100 scale to the ranked items by considering each lower item. This procedure converts the ranked items into weighted items on the interval scale according to the relative importance. In contrast, the pair-comparison method provides a more accurate and discriminatory ranking technique to judge the priority order of the list of attributes by comparing the attributes in pairs compared to the arbitrary ranking methods. In regular pair-comparisons the items are compared in pairs in a matrix consisting of all the listed items until all the listed pairs are completed and the important item is considered in the pair. The frequency along the row for each item is then computed to determine the importance of the attributes. In scaled-pair comparison a preference weight is assigned to each comparison along with the ranked item and the total row of normalised rating is computed to determine the rating of the attributes. Conversely, in pair-cost-comparison real monetary amounts are assigned along with the preferred item depending on the cost differentiation. Direct magnitude estimation is a technique used in a group decision-making environment to assign suitable weights along the principles described in earlier techniques by computing the geometric mean for each item. Whereas in category scaling a finite scale range is utilised according to the preferences for the category description in a group decision-making situation. Criteria analysis provides a procedure for quantification by systematically multiplying the weight of every criterion by the rating of the functions, which are dependent on the perception of the contribution of each component of the functions. The scoring model is similar to criteria analysis; nevertheless it provides a comprehensive mechanism to evaluate the alternatives by comparing the performance measure scale established for each criterion.

None of the above methods assign any monetary value for an attribute except the pair-cost-comparison method. This method utilises monetary units to express the difference between two items, in a pair of items being compared. However the Cost-Value method proposed by Joslin (1968) provides a framework to assign monetary values to intangibles without resorting to pair comparison among intangibles. On the other hand, when an AMT is considered for investment, comparing the intangible and tangible attributes in a pair is an irrelevant exercise.

3.6.1 Cost-Value Technique

The Cost-Value technique was originally proposed to evaluate the selection of computers among vendor proposals. However, this Cost-Value technique can be utilised to evaluate the acquisition of AMTs from vendor proposals. The application of the Cost-Value technique enables the decision-maker to understand the importance of the attributes in the evaluation process and helps to make informed decisions.

In traditional value analysis techniques, the value is considered consisting of the following elements:

1. Esteem value: The qualities of the product which give desirability and respect;
2. Exchange value: The worth of the product which is transferable or exchangeable; and
3. Utility value: The functions of the product which are in need and useful.

Then the value is expressed as

$$\text{value} = \frac{\text{function}}{\text{cost}} = \left(\frac{\text{esteem} + \text{exchange} + \text{utility}}{\text{cost}} \right) \tag{3.5}$$

English dictionaries define the value as "material and monetary worth". Naturally the value of any item is proportional to the perceived benefits derived from that item. By dividing the function by cost to derive the value in the traditional equation we are diluting the meaning of the value to a dimensionless index. Hence in this research the value is considered as the perceived benefits of an attribute associated with implementing an AMT.

The Cost-Value technique involves evaluating desirable and valuable features and assigning a cost for them. This technique combines the simplicity of the cost only technique, where only the cheapest item is considered for consideration with a weighted scoring technique. The Cost-Value technique requires the required features of the system considered, to be categorised as mandatory and desirable features. Then the technique enables the decision-maker to evaluate only the desirable features of a given system and the cost of it. The mandatory features are not evaluated and these features should be provided in the system for the initial consideration. By this approach the additional and desirable features above the mandatory features offered by the vendors are also appraised. The important feature of the Cost-Value technique is assignment of monetary value to the desirable features. The decision-maker appraises any additional features to ascertain whether they deserve to be included in the evaluation process, and if so, determines the Cost-Value. This enables the decision-makers to realistically distinguish the products offered by various vendors with a common parameter.

3.7 CONCLUDING REMARKS

In this chapter we have briefly identified justification approaches for CIM such as economic, analytic and strategic approaches. Later, we have realised that multicriteria techniques should be utilised to evaluate multiple goals of a company. The multicriteria techniques of AHP and the weighted scoring are useful in prioritising the strategic objectives of the company. In addition, we have seen the fundamentals of the economic tools and the basis of the Cost-Value technique. Having identified these tools, we need to look forward for optimisation techniques and integrating them into a single strategy. Therefore, the next chapter is devoted to optimisation techniques for CIM investments.

3.8 REFERENCES

1. Boaden, R.J. and Dale, B.G., 1988, "Justification of computer integrated manufacturing: some current trends", *International Journal of Computer Integrated Manufacturing Systems*, Vol 1 No 4, pp. 198–203.
2. Boddy, D. and Buchanan, D.A., 1986, *Managing New Technology*, (New York: Blackwell).
3. Hutchinson, G.K., 1976, "Production capacity: CAM Vs transfer line", *Industrial Engineering*, Vol 8 No 9, pp. 30–35.
4. Joslin, E.O., 1968, *Computer Selection*, (USA: Addison-Wesley Publishing).
5. Kaplan, R.S., 1986, "Must CIM be justified by faith alone", *Harvard Business Review*, Vol 64 No 2, pp. 87–95.
6. Krinsky, I. and Miltenburg, J., 1991, "Alternate method for the justification of advanced manufacturing technologies", *International Journal of Production Research*, Vol 29 No 5, pp. 997–1015.
7. McKay, M. and Galbraith, J., 1989, "Financial Justification of Advanced Manufacturing Technology Through Strategy Qualification", *Proceedings of 5th National Conference on Production Research*, pp. 123–127.

8. Meredith, J.R. and Suresh, N.C., 1986, "Justification techniques for advanced manufacturing technologies", *International Journal of Production Research*, Vol 24 No 5, pp. 1043–1057.
9. Mohanty, R.P. and Venkatraman, S., 1993, "Justification study for computer integrated manufacturing", *International Journal of Computer Integrated Manufacturing*, Vol 6 No 6, pp. 366–374.
10. Radcliff, B., 1986, "Multi-criteria decision-making: a survey of software", *Social Science Microcomputer Review*, Vol 4 No 1, pp. 39–55.
11. Saaty, L.T., 1980, *The Analytic Hierarchy Process: Planning, Priority Setting, Resource Allocation*, (New York: McGraw-Hill International).
12. Sullivan, W.G., 1986, "Models IES can use to include strategic non-monetary factors in automation decisions, *Industrial Engineering*", pp. 42–50.
13. Weber, S.F., 1992, "Multi-criteria decisions on automation investments", *Proceedings of Manufacturing International 1992*, Texas, pp. 159–167.

CHAPTER FOUR

Integrated Approach to CIM Justification and Optimisation

4.1 OVERVIEW

Some commercial companies grow, improve their market share, prosper and become profitable while fulfilling their goals, but other companies dwindle, lose their customer base, become non competitive and bankrupt or taken over by growing companies. Why do the managers of the companies let this scenario happen? It is because there is a difference in formulating strategies to run the operation of the company both in successful and in failure mode. The proposed integrated justification and optimisation approach discussed in this chapter provides an answer to propel the companies towards the success mode by investing in advanced technologies. The integrated approach effectively utilises widely used multicriteria models such as the weighted scoring technique and analytic hierarchy procedure (AHP), the traditional financial techniques and the Cost-Value method, all of which were discussed in Chapter three.

This chapter and following two chapters present a systematic and practical approach in relating strategic consideration of the enterprise for selecting suitable advanced manufacturing technologies (AMTs) and assigning them in an optimal manner. The primary focus of this chapter is to provide a selection mechanism for identifying suitable AMTs as components of CIM in an integrated approach to CIM justification and optimisation. This chapter opens with a discussion on the need for strategy based decisions and explains a methodology for an integrated justification and optimisation of CIM.

4.2 INCORPORATING STRATEGIC FOCUS INTO CIM JUSTIFICATION

Manufacturing companies are constantly striving to improve their competitive capability by investing in advanced technologies in today's international and local competition. AMTs have been identified as tools which can provide that competitive capability. However, existing traditional economic analyses do not have a facility to incorporate strategic requirements to choose suitable AMTs. Generally, manufacturing managers have a predisposition regarding which technology is to be selected for their company. This predisposition poses a potential danger in selecting subsystems which should not have been selected at all. In addition, the available literature does not contain a clear conceptual model that links the strategy, selection of technology and existing knowledge in a cohesive manner.

Manufacturing companies are looking for directions to improve their performance, compared to their competitors, by investing in AMTs. Improving the performance of a company by achieving technological competitiveness is a necessary condition to ensure market competitiveness, otherwise the company will be fighting a loosing battle. Competitiveness is derived from many factors including increased productivity, being responsive, proactive and innovative, quality, flexibility, and reduced inventory. Selection of AMTs to increase the competitiveness, based on optimal resource allocation, offers a unique challenge to manufacturing managers, since the selection and optimisation process to achieve the strategic objectives is a complex process because of many trade-offs among conflicting factors and it has serious implications. In addition, the selection and optimal allocation process should consider tangible and intangible factors associated with each AMT and the technical complexity of the equipment. Traditionally the selection of equipment relied on assessing the financial return by the investment on that equipment and it is considered only as a stand-alone investment disregarding functional inter-relationships and attributes associated with it.

The reluctance to implement in available innovative technologies and/or the analysis of implementation decisions within a limited scope has resulted in a reduced level of competitiveness for many manufacturing industries (Chen and Small, 1994). Besides, a clear consensus has not yet been achieved to address CIM investments. Several researchers and consultants suggest various methods of approaching this issue. Strategic decisions are the milestone of any successful industry. An analytical approach helps in identifying the issues surrounding a problem in a logical and easily understandable manner. The economic benefit from any investment is the bottom line of any new proposal for a company. Meredith (1987) stated the importance of a comprehensive justification approach that can conceptually handle quantitative and qualitative factors together. By combining strategic, analytic and economical approaches in an appropriate manner, a comprehensive analysis can be performed on CIM/AMT investments and a better decision with respect to justification can be achieved.

In order to accommodate the above view, an approach that incorporates all the required evaluation techniques for justification has been developed. Furthermore, this approach simplifies the process of analysing by guiding the manufacturing manager through the justification process in a sequential procedure that consists of six stages. Accordingly, this approach has been formulated as a multi-attribute spreadsheet model and developed as a decision support system (DSS) for CIM investment analysis (DSSCIM). The DSSCIM allows the decision-maker to explore various multiple alternatives by an interactive program and to make the best decision under the given constraints. The DSSCIM includes most of the factors that may affect the investment decision. The complete flow chart of the DSSCIM is shown in Figure 4.1.

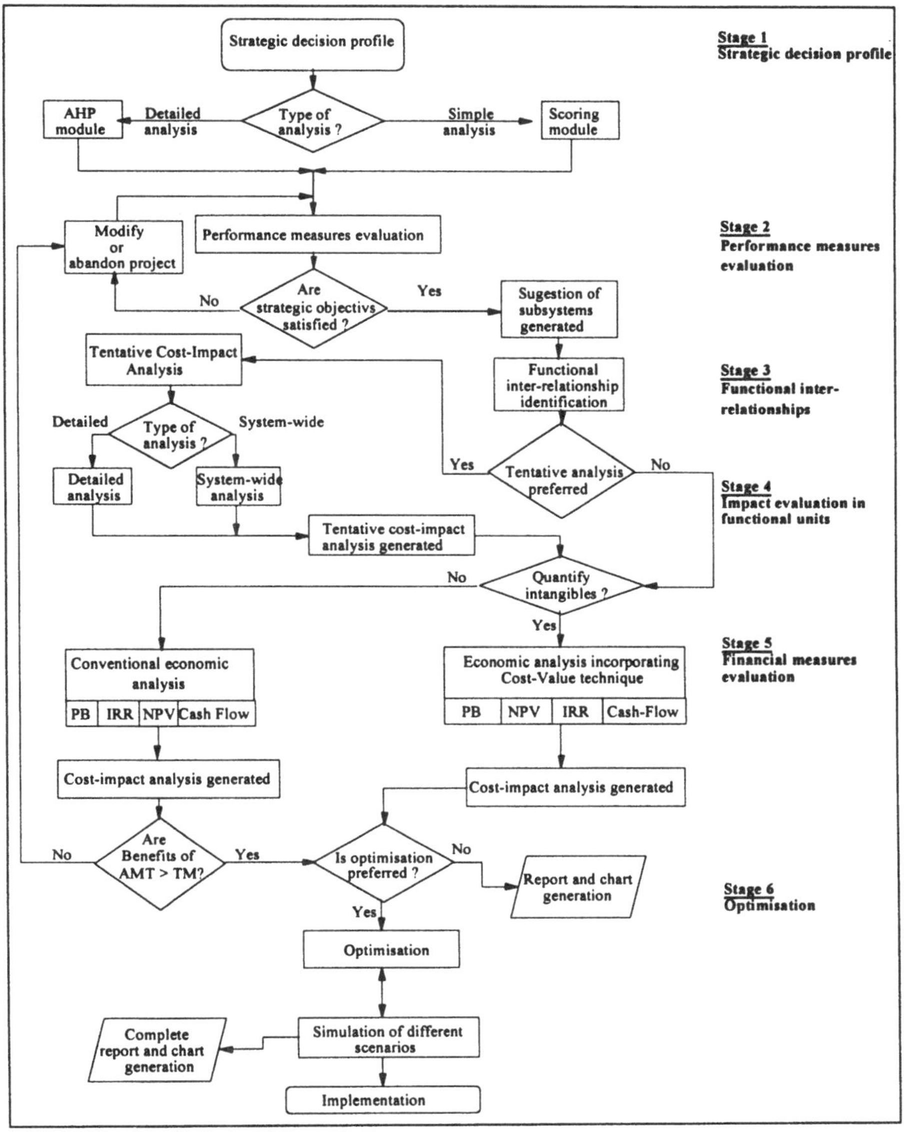

Figure 4.1 Flow chart of the DSSCIM

4.3 THE METHODOLOGY

In an increasingly competitive world, every manufacturing company is trying to capture more market share by continuously improving their products. The market share depends on the ability of the competitors in the market. It may decrease, instead of staying at the same level, if investments are not made towards AMTs, since we cannot expect the competitors to stay at the same level. Integrated manufacturing provides a capability to compete locally and globally as long as the systems chosen are able to provide a competitive edge in cost, delivery, quality and responsiveness in an optimal manner. In a usual decision-making practice the type of technology is already pre-selected by the managers depending on their preferences for the particular AMTs when it is decided to evaluate the possibility of introducing AMTs in a company (Primrose, 1991). Having selected the AMT, the managers strive to justify by direct savings and if it cannot be achieved they venture towards identifying the benefits by the intangibles (Primrose, 1991). Attempting to justify investments in this particular manner leads to difficulties in justifying the investment in AMTs and results in technologies which are not consistent with the requirements and resources of the company. In addition, this selection procedure will result in technologies which are not the optimal solutions for the company.

Identifying appropriate technologies in the face of the plethora of technologies available, requires a considerable knowledge in decision-making and a structured approach in refining the search for the technologies. Keeping these concerns in mind the following methodology is proposed. The proposed methodology explained in this chapter consists of six sequential stages of analysis. The stages are designed according to the requirements of data input. However, the sixth stage can be performed independently of the fourth and fifth stages. The stages in the integrated justification and optimisation approach are:

- **Stage 1: Strategic decision profile**
  This stage involves identifying the strategy of an enterprise. The strategy is identified by selecting attributes relevant to AMTs. Then the priorities of the selected attributes are determined either by the AHP or by the weighted scoring method.
- **Stage 2: Performance measures evaluation**
  Performance measures evaluation stage focuses on evaluating selected attributes in the strategic decision profile with regard to the option of investing in AMT or in conventional machinery. This stage enables the decision-maker to identify the expected goals of an investment in AMTs. From the expected performance measures suitable subsystems can be identified with knowledge available from companies which have already installed AMTs. If there is a need to analyse the selection of AMTs thoroughly, then a rule-based algorithm can be formulated.
- **Stage 3: Functional inter-relationship identification**
  The prime benefit of investing in an AMT is that it brings benefits not only to the primary functional unit where the technology had been installed but also to the other functional units of the company. Hence, this stage involves identifying the synergy among technologies and various functional units of a company by a systematic procedure or by estimation.
- **Stage 4: Impact evaluation in functional units**
  Impact in functional unit evaluation enables the decision-maker to investigate the effect of an investment in various functional units and on the company as a whole before proceeding to the actual investment and implementation. This

procedure allows computing system-wide and departmental-wide tentative cost-impacts of investment from the estimated benefits of attributes related to each identified technology.

- **Stage 5: Financial measure evaluation**
  This stage involves calculating the financial measures related to an investment and evaluating the investment with the tangible and intangible attributes of AMTs. At this stage, the Cost-Value technique is utilised for quantifying the intangibles. However, the users are given an option of not using the Cost-Value technique as well.
- **Stage 6: Optimisation**
  This stage provides an optimal resource allocation to satisfy the strategic requirements of an enterprise. A goal programming (GP) algorithm is utilised to build the relationships between the attributes and the chosen subsystems to determine the best combination of AMTs for the CIM environment to achieve the strategic objectives. The GP optimisation approach is explained in the previous chapter. In addition, this stage provides facilities to incorporate the synergy and the limitation of the capital into the optimisation procedure.

The technologies considered as examples in the proposed methodology in this book are; Office automation, CAD, CAM, CAPP, CAQC, AS/RS, robotics, GT, CM, FMS. The additional CIM technologies, which can also be considered in this type of methodology, are numerous and some of which are mentioned in Chapter one, Section 1.2.1.

While many of the CIM technologies have been around for some time, many companies have experienced and are experiencing problems associated with CIM implementation due to improper planning and by failing to recognise the strategic benefits of CIM. Improper planning will result in the selection of inappropriate technologies to support the corporate objective. The significance of the proposed approach in this book and the existing approaches are:

1. The proposed approach provides a unified approach, which enables strategic, analytic and financial analyses in a single methodology and combines approaches for quantifying intangibles and indirect costs;
2. In the proposed approach the AHP is not directly utilised to evaluate the technologies for the strategic attributes as it has been by earlier researchers, but a database and rule-based algorithm is used to generate the suitable preliminary subsystems. This feature avoids the bias towards the technologies by the manufacturing managers and strengthens the decision-making process; and
3. The proposed methodology enables evaluating impact of the subsystem in the company and utilises synergy among various technologies to find an optimal resource allocation on the selected technologies.

4.3.1 Stage 1: Strategic Decision Profile

The absence of reliable, accurate and easy to use criteria, which reflects the strategic benefits of the company, and the measures of the effectiveness and efficiency of the company has been stated as one of the important factors holding companies in investing in AMT (Berliner, and Brimson, 1988). To be effective in selecting components of CIM related to the strategic objective and be realistic in analysing the investment, out of the intangible and tangible attributes available for AMT/CIM, only twenty two important attributes had been identified and suggested to be used in the methodology. However, the users of the methodology can choose some other attributes other than suggested, if they prefer them. The present selection is made with regard to the importance of the attributes and availability of information related to attributes in literature. The selected attributes are categorised under five classes according to their feature for clear understanding and analysis. When using a pairwise comparison process in AHP methodology, it has the limitation of requiring the pairwise comparison to be made among factors which do not have a common basis for comparison, such as "how important is increasing product growth than reducing personnel". However, when we group similar attributes, which have a common base for comparison, in a category, the limitation is avoided. The five suggested categories and their attributes are shown in Table 4.1.

Table 4.1 Grouping of attributes

Cost Reduction Factors	
Direct	**Indirect**
Direct labour (reduction)	Maintenance Costs (reduction)
Material Costs (reduction)	Production planning & control (reduction)
Material handling Costs (reduction)	Floor space (reduction)
Tooling/Fixture Costs (reduction)	Inspection (reduction)
Scrap (reduction)	

Productivity Improvement Factors	
Production related	**Human related**
Manufacturing productivity (increase)	Design flexibility (increase)
Machine utilisation (improve)	Production flexibility (increase)
Work in progress (reduction)	working conditions (improve)
Rework (reduction)	Reduce paperwork (reduction)
Lead-time (reduction)	

Business Strategy Related Factors
Increase product growth
Reduction in delivery times
Ability to introduce new products
Ability of the system to expand

The grouping of attributes offers a facility to assign importance of one category over another in a strategic decision concept or to rank these categories in a suitable order. Initially decision-makers are advised to use all the offered attributes. By this action, we will be able to carefully consider the attributes which are strategically desirable for the company. The selection of attributes can either be qualitative or quantitative, but the attributes must be strategically important to the company. Having done the selection of attributes, we are ready to proceed to the analysis of succeeding stages.

4.3.2 Stage 2: Performance Measure Evaluation

Performance measurement is a measure of the expected improvement in business activities by AMT/CIM, in relation to the specific goals and objectives developed in the strategic planning process. To be effective, we should establish the performance measures in congruence with the objectives of a company. Many manufacturing companies fall into difficulties by evaluating the performance of the manufacturing facilities on the basis of only cost and efficiency (Skinner, 1985). In any investment which has strategic implications for the organisation, identifying the required performance measures on most of the criteria is essential and should be an integral part of any investment procedure. Especially for AMT, a clear understanding in the performance measures expected will make it excellent for an appropriate subsystem or systems identification, which will bring the perceived benefits to the company. In addition, by setting a clear concept in performance goals, we will be able to verify the acceptance of CIM into the corporate strategy in our justification and optimisation process.

In today's environment manufacturing management must change the emphasis on the cost-based criteria as the sole indicator of manufacturing performance to more meaningful multi-attribute criteria to realise the benefits of AMTs. However, we need to make sure that the multicriteria performance measurement system is capable of incorporating both efficiency and effectiveness. To provide effectiveness in measuring the performance measurement indicator and to improve the realisation of benefits assessment, a multi-attribute performance measurement system is proposed in this chapter. The performance measurement can be performed by either the option of AHP or simple ranking provided in this methodology. However, AHP enables a detailed analysis while simple ranking provides a simplistic performance measure evaluation method.

4.3.2.1 Analytic hierarchy process

AHP provides a comprehensive structure to combine and evaluate the rational and irrational values in a decision-making process. In the proposed methodology the hierarchy of the problem to evaluate the strategic importance of the decision is structured in following three levels:

- At level 1, the categories are compared and ranked, with respect to the overall strategic objective of the company;
- At level 2, the attributes within the criteria are compared and ranked against each other; and
- At level 3, the weight of AMT and the Traditional Machinery (TM) is verified, with respect to the attributes on level 2.

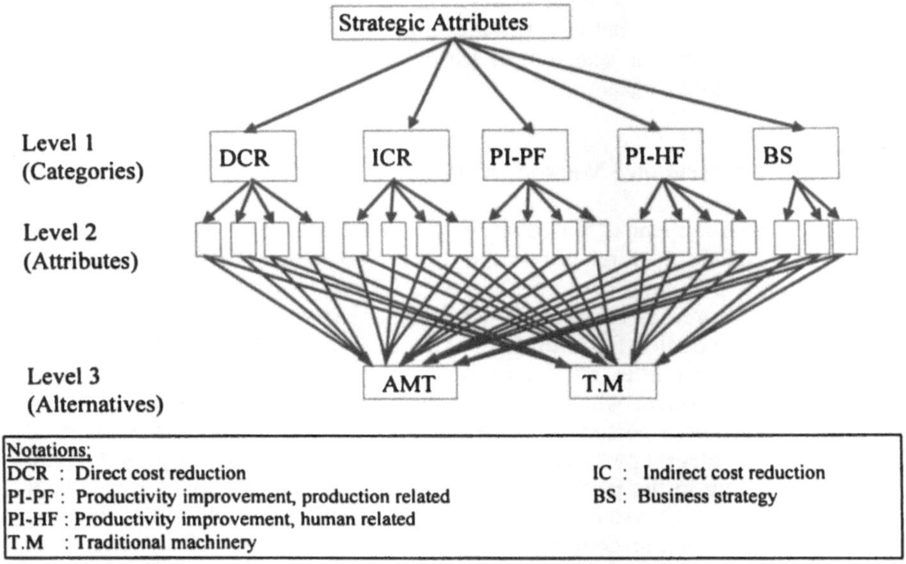

Figure 4.2 Three levels of analytic hierarchy process

The prioritisation of attributes for level 1 and level 2 can be performed by the decision-maker by assigning values on the subjective quantitative scales 1 to 9. The procedure of assigning values on this scale has been highly recommended by Saaty (1980) for a viable assessment in any pairwise comparison process. The method of using AHP is explained in Chapter three. In each cell of the pairwise comparison matrix, the decision-maker indicates how many times as important the attribute in each row is compared with attributes in each column. The values of the upper triangle for the comparison matrix is to be completed in this procedure by the decision-makers and the remaining values are the reciprocals of the values in the upper triangle. The AHP methodology provides a mechanism to calibrate the consistency index to improve the judgements for the overall consistency in comparative judgement. The index is the deviation from consistency and the upper limit of that index specified by the AHP is 0.10. The method of consistency index computation is explained in Appendix 1.

In this procedure, we need to calculate the consistency index for the comparison matrix and verify whether it is less than 0.10. If it is found to be otherwise, we may reassign ranking to evaluate again. However, if the priority ranking is satisfactory in our opinion, we can override this verification. On completion of pairwise comparison process at level 1, we continue the pairwise comparison process at level 2, for the attributes in each category. By completing the AHP procedure for all the attributes in all the categories, then we will be in a better position to identify the importance of the qualitative and quantitative factors which are desirable to the company. On completion of the pairwise comparison for all the categories at level 2, we need to transfer prioritised weights of each category to a performance measures evaluation sheet. This procedure enables the consistency of prioritisation to be achieved, among all the selected attributes. Then the final stage of the AHP pairwise comparison procedure at level 3 can be conducted on a performance evaluation sheet.

4.3.2.2 Weighted scoring method

The weighted scoring approach involves quantifying the relative importance of each attribute by assigning a suitable weight for each attribute. By using the weighted scoring option decision-makers will be able to analyse the attributes without resorting to pairwise comparison procedures. This option can be used by a decision-maker when weights can be comprehensively assigned to the attributes from experience or when sufficient information is available for this procedure. In the weighted scoring module option, the performance measures of investing in AMT is compared with status quo. Although the status quo is not an ideal base to compare the strategic decision which affects the company, it can be used to measure the impact of not investing in AMT, and can force the decision-maker to understand the nature of AMT investment. If the investment is a green-field investment, the comparison can be evaluated between investing in TM and AMTs. A typical estimate of the performance analysis by the simple weighted scoring module is shown in Table 4.2.

For this typical estimate the weights and ranking have to be supplied by the decision-maker. Then the feasibility of the investment can be estimated. However, many of the tasks need to be performed by the decision-makers are automated in the DSSCIM developed by the authors of this book. If necessary, as a help tool we will be able to use sample weights, which are gathered from available literature. The sample weights, which are included in the database, are given in Appendix 2.

Table 4.2 A typical performance measure evaluation by the weighted scoring method

Attribute /Criterion	Weight	Ranking AMT	Ranking Status Quo
Direct Cost Reduction			
1 Direct Labour	20.0	7.00	1.00
2 Material Costs	15.0	6.00	1.00
3 Material Handling Costs	15.0	5.00	1.00
4 Tooling/Fixture Costs	18.0	4.00	1.00
5 Scrap	35.0	7.00	1.00
Indirect Cost Reduction			
1 Maintenance Cost	28.0	1.00	4.00
2 Prod. planning & control	30.0	6.00	1.00
3 Floor Space	20.0	5.00	1.00
4 Inspection	40.0	8.00	1.00

Productivity Improvement (Production Factors)				
1	Manufacturing productivity	68.0	8.00	1.00
2	Machine utilisation	46.0	9.00	1.00
3	Work in progress	30.0	6.00	1.00
4	Rework	20.0	5.00	1.00
5	Lead-time	30.0	4.00	1.00
Productivity Improvement (Human Related Factors)				
1	Design flexibility	50.0	4.00	1.00
2	Production flexibility	50.0	6.00	1.00
3	Working conditions	30.0	6.00	1.00
4	Paperwork	10.0	3.00	1.00
Business Strategy				
1	Product growth	70.0	6.00	1.00
2	Delivery times (Reduction)	75.0	6.00	1.00
3	Introduce new products	45.0	4.00	1.00
4	System expansion	25.0	5.00	1.00

4.3.3 Comparison of Performance Measures

The comparison of performance measures enables the decision-maker to establish the justification of CIM/AMT on the strategic level. The procedure of evaluating performance measures can be handled by either of two methods described in previous section.

For the AHP method, this procedure is handled as a comparison process of two alternatives. For the AHP method; let

a_{amtj} = Priority weight (principal vector) of an attribute for AMT at level 3 (j = 1 to n)
a_{tmj} = Priority weight (principal vector) of an attribute for TM at level 3
C_i = Priority weight (principal vector) of category at level 1 (i = 1 to 5)
$T_{amt,i}$ = Vector multiplication for the i th category of AMT
$T_{tm,i}$ = Vector multiplication for the i th category of TM
W_j = Priority weight (principal vector) of an attribute at level 2

As shown in Appendix 1, when there are only two decision parameters such as AMT and TM to be compared over another, and if the preference of an attribute for AMT is X times more than the other attribute for TM, the priority weight can be computed as:

Priority weight of the j th attribute for AMT $= a_{amtj} = X/(X+1)$
Priority weight of the j th attribute for TM $= a_{tmj} = 1/(X+1)$
Utilising above notations, the vector multiplication of the priority weights can be written as:

For AMT
$$T_{amt,i} = C_i * \sum_{j=1}^{n} W_j * a_{amt,j} \tag{4.1}$$

Similarly, for TM
$$T_{tm,i} = C_i * \sum_{j=1}^{n} W_j * a_{tm,j} \tag{4.2}$$

When
$$\sum_{i=1}^{5} T_{amt,i} > \sum_{i=1}^{5} T_{tm,i} \tag{4.3}$$

is achieved, the justification on the strategic level is satisfactory, since cumulative weights for the attributes of AMT are greater than the cumulative weights for the attributes of TM. Then we can proceed to the next stage of analysis.

For the weighted scoring module; each attribute is weighted by multiplying the evaluation by its corresponding normalised criterion weight. The overall score is then determined by summing its weighted evaluations with respect to all the criteria; let

I_i = Weight of a criterion (i = 1 to 5)
P_{aj} = Performance level for AMT investment for an attribute (j = 1 to m)
P_{sj} = Performance level for Status Quo for an attribute
$T_{amt,i}$ = Total scoring for the i th category of AMT
$T_{s,i}$ = Total scoring for the i th category of Status Quo
W_j = Weight of an attribute

Utilising the above notations, the total multiplication of the weights can be written as:

For AMT
$$T_{amt,i} = I_i * \sum_{j=1}^{n} W_j * P_{aj} \tag{4.4}$$

Similarly, for status quo
$$T_{s,i} = I_i * \sum_{j=1}^{n} W_j * P_{sj} \tag{4.5}$$

If
$$\sum_{i=1}^{5} T_{amt,i} > \sum_{i=1}^{5} T_{s,i} \tag{4.6}$$

then it is clear that the cumulative weight for investment AMT is higher than the cumulative weight of continuing with conventional technologies or investing in conventional machinery. When this is achieved, according to the proposed approach we can proceed to the next stage.

4.3.4 Identifying Suitable Subsystems

The selection of appropriate technology and acquisition decision is the major function of the decisions-maker. Generally, most decisions are made by mental models, which rely on the knowledge, experience, intelligence, and biases of the decision-maker. The knowledge required to make a correct decision can be gained in following ways:

1. Gathering information related to AMT and its implementation cost from the vendor literature, by reviewing relevant publications. This information gathering can be enhanced with gathering detailed specification, and relevant warranty, service and spare parts support from vendors, and by administrating questionnaires; and
2. Visiting other companies which already have the AMTs and determining its performance, reliability, inter-functional data input/output for the company concerned by observing and analysing, by inspecting records and documents, and by interviewing the decision-makers and people who have implemented CIM in that company, etc.

Each company has different needs and operates in different circumstances. Hence, acquired knowledge must be carefully analysed before considering a possible application for the particular company. Selecting the appropriate subsystems from available technology to achieve the same strategic objective is a complex issue for decision-makers, due to the variety of technologies available in the market. The proposed methodology reduces the complexity associated with selecting a subsystem by providing suitable suggestions based on the requirement. The use of a strategic and analytic approach to find appropriate subsystems for the company reduces the bias of the decision-maker in forming the solution and provides a constructive approach to find an answer.

To identify suitable subsystems, each subsystem considered for investment needs to be associated with attributes which are offered at the strategic decision profile. This association can be carried out depending on the feature of the attribute and on the feature of the subsystem, based on available literature on AMT/CIM. The associated attributes for each technology considered in the methodology are given in Appendix 2, which can be used by decision-makers as a guide for identifying attributes and technologies. However, the DSSCIM developed by the authors automatically generates suggested subsystems for the investment by analysing weight and ranking provided on the performance-measures-state and by using a rule-based algorithm.

The proposed methodology and the algorithm developed by the authors of this book try to answer the perplexing question: what technologies are necessary to meet the company's business objectives? By utilising Equations 4.7 to 4.9, suitable subsystems can be identified for further analysis.

When AHP procedure is utilised; let

W_j^{T*} = Weight of an associated attribute for technology T_i at level 2

$T_{amt,i}^{T*}$ - Vector multiplication of associated attributes (of technology T_i) for the ith category

In addition to the above notations, by incorporating the notations given in Section 4.3.3, the Total Weight of Associated Attributes of T_i among selected attributes (TWAA of T_i) can be written as:

$$\text{TWAA of } T_i = \sum_{i=1}^{5} T_{amt,i}^{T^*} \tag{4.7}$$

$$\text{Where } T_{amt,i}^{T^*} = C_i * \sum_{j=1}^{n} W_j * a_{amt,j} \in T_i$$

When a simple scoring procedure is utilised; let

$W_j^{T^*}$ = Weight of an associated attribute for technology T_i

$T_{amt,i}^{T^*}$ - Total scoring of associated attributes (of technology T_i) for the ith category

In addition to the notations given above, by incorporating the notations given in Section 4.3.3, the Total Weight of Associated Attributes of T_i among selected attributes (TWAA of T_i) can be written as:

$$\text{TWAA of } T_i = \sum_{i=1}^{5} T_{amt,i}^{T^*} \tag{4.8}$$

$$\text{Where } T_{amt,i}^{T^*} = I_i * \sum_{j=1}^{n} W_j * P_{aj} \in T_i$$

When TWAA of T_i is used as an indication, it will highlight the cumulative weights for an AMT within the strategic decision profile. Once all the values are obtained, the TWAA of T_i can be depicted in a histogram as in Figure 4.3. The figure shows the results for the weights and ranks listed in Table 4.2 (page 61), with equal weights assigned to the categories.

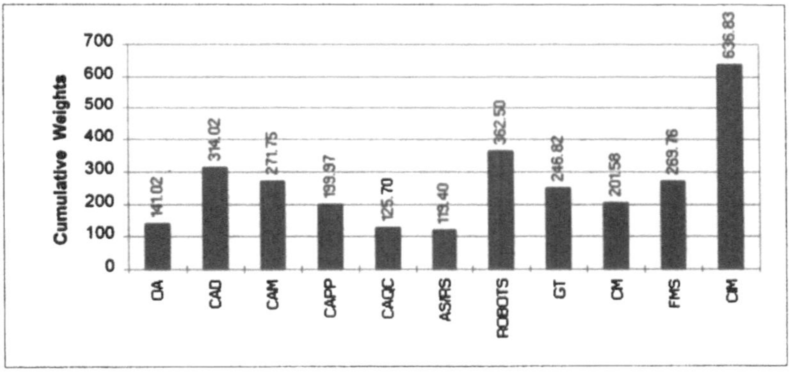

Figure 4.3 Histogram for the cumulative weights of associated attributes

The use of a histogram will assist the decision-makers to select suitable subsystems appropriate to the company in a structured manner. The decision-maker can utilise the results or depending on the preference can choose alternative subsystems for further analysis. If we like to verify the expected level of an AMT with the average performance levels for the AMT with the help of the database, we could compute the performance ratios as in Equation 4.9.

For Technology T_i, performance ratio

$$= \frac{\text{Total weight of associated attributes of } T_i \text{ among selected attributes}}{\text{Total weight of associated attributes for } T_i \text{ from Database}} \quad (4.9)$$

A typical chart for the weights and ranks listed in Table 4.2 (page 61), with equal weights assigned to the categories is shown in Figure 4.4. The weight of the histogram indicates ratio of the expectation for an AMT with the average performance levels from the database.

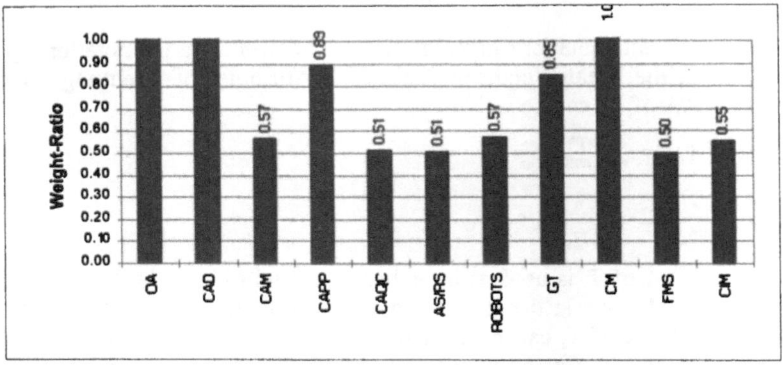

Figure 4.4 Histogram for relative ratio of subsystems

There is no standard path towards implementing CIM environment and it may vary depending on the requirement of a company and the position the company holds in the industry. However, some companies have become leaders in the field of CIM. Learning from them would be a valuable experience for a manufacturing manager who is interested in implementing CIM for his/her company. The methodology proposed in this book tries to fill the gap in experience and knowledge by utilising a database, obtained from leaders in CIM /AMT implementation. However, to make the database be more effective, further research should be conducted to understand the aspects of integration and gains in using CIM in a company. Then the collected information can be very useful to manufacturing managers to apply in the proposed approach.

4.3.4.1 Limitations of the database

Data gathered from the literature survey will generally have limitations placed on them, since we do not have a control on the validity and accuracy of the data gathered. The data from the literature is entirely dependent on the objective and the background of the literature. In addition, the willingness of the companies to divulge real information is also a difficult aspect to measure unless everyone concerned fully participates in developing a comprehensive and actual database. Accurate information related to performance improvement can be gathered by focusing on the attributes considered in this book by an extensive questionnaire, and by examining the data provided by the companies in a thorough manner. Some of the attributes obtained from the literature have a large range of values and various authors have provided various ranges for the same attributes, so a median value is assumed. Furthermore, the measurement of an attribute is dependent on the definition considered by the researcher for that attribute. These aspects also place limitation on the accuracy of the database. In addition, the literature does not provide values for a few attributes considered in this database, so some reasonable values are assumed. Nevertheless, the expected percentage values do not diffuse the concept of selecting appropriate technologies, since the selection of appropriate technologies are not dependent on them but on the associated attributes for each AMT. However, the methodology proposed in this chapter illustrates the usefulness and easiness in selecting appropriate subsystems when a good database is available to the manufacturing managers.

4.3.5 Equipment Selection Procedure

The selection of equipment including hardware and software components for automation depends on many different parameters including cost of hardware and software, cost of peripheral equipment, after-sales vendor support, size of the plant, available space for expansion, site preparation cost, cost of direct and indirect labour, costs and benefits of intangibles, type of production such as mass, batch and job, production rate, complexity of the product, available automation technology to meet the demand and desired degree of automation. Furthermore, the functional inter-relationships with identified technology with respect to other AMTs and the functional units have to be evaluated for proper equipment selection. Since, as stated earlier, the benefits from AMTs will be greater with integrated applications than from an unlinked or individual application.

Selecting and short-listing of suitable vendors to provide the required equipment should be carried out simultaneously with the identification of suitable technologies. The availability of after-sales vendor support is crucial in identifying the appropriate vendors in some countries, since some vendors in this complicated market have a possibility to become bankrupt. In addition, the variety, quality, and quantity of the after sales support also should be taken into consideration to identify and short-list appropriate vendors among the available group. However, the quality and the quantity of the after sale support required by the company depends on the technical ability and capacity of the company concerned as well. Weatheral (1992) recommends a simple scoring method to choose the suitable vendors. The required characteristics for the vendor performance may consist of features including, price of the equipment, implementation cost if any, delivery lead-time, lead-time for maintenance support, quality of the software and hardware, availability of the peripheral equipment, and the potential of the vendor to provide future needs (Weatheral 1992). This procedure requires an extensive knowledge about all the potential vendors, who could supply the equipment. Boaden and Dale (1990) proposed five stages for the vendor contact. They are:

1. Explanation of the strategy and presentation of technical specification to vendors. At this stage, the vendors are supplied with adequate information on

the strategically required CIM components for the company to achieve the CIM strategy and the detailed specifications of the required subsystems;
2. Demonstration of system and broad costing. Once the detailed specifications are given to the vendors, the demonstration of the available equipment at the vendor's site has to be arranged. The vendor is requested to provide the quotation and broad-outline costing for the equipment before the demonstration is conducted;
3. User visits. At this stage, the users of the system are taken to the vendor's site to give their opinion on the equipment;
4. Detailed quotation. A detailed quotation including; the cost of the software and hardware, integration and installation costs, and consultancy cost is requested from the vendor; and
5. System weighting. Once participants of the decision-making group have seen the demonstration, the participants are requested to provide the weighting for requested parameters.

Once the technologies have been identified to satisfy the requirement of an enterprise, the variety of equipment available under the same technology from the chosen vendors can be evaluated by the scoring model provided by Boaden and Dale (1990) or by the Cost-Value technique proposed by Bhat (1992). The use of the Cost-Value technique is far superior to any other evaluation technique available. This selected equipment can be comprehensively analysed by the financial modules explained in this chapter. The utilisation of the unified approach provides a comprehensive analysis for selection of various AMTs for the integrated manufacturing environment, by incorporating the Cost-Value technique, which we have identified in Chapter three, Section 3.7.1. On completion of the preliminary selection of subsystems, the integration aspects and the determination of synergy among integrated technologies plays a vital role in choosing the appropriate combination of subsystem for CIM environment to satisfy the strategic objectives.

The effect of synergy due to the integration of technologies will have an impact on various functional units of an enterprise. Chapter two provides theoretical development and guidelines to quantify the functional inter-relationships in an enterprise towards the integration of AMTs. In addition, a mathematical model to evaluate synergy among interconnected technologies is also proposed in Chapter two. The derived functional relationships are utilised in association with a synergy model in the methodology and the DSSCIM in evaluating the impact of investing in AMTs in an enterprise under the framework of CIM and realising the optimal investments for subsystems of CIM. The impact evaluation stage is discussed in Chapter five.

4.7 CONCLUDING REMARKS

This chapter presents an approach, where the strategic concerns of the company is identified, performance measures for the identified strategy are determined, and suitable subsystems are determined based on the expected performance measures. Most importantly, this chapter illustrates a methodology to select AMTs which fulfil the strategic concerns of a company. Identifying suitable technologies based on the strategic consideration of the company will narrow down the selection of appropriate AMTs among the plethora of available technologies. The identified AMTs can be later utilised based on an optimal resource allocation approach, which is discussed in Chapter six.

The next chapter describes the impact and financial evaluation analysis of the proposed integrated approach to CIM justification and optimisation. The integrated approach, which utilises multicriteria decision-making principles, will help decision-makers who are willing to invest in CIM. The integrated approach is incorporated into a decision support system for CIM justification and optimisation (DSSCIM). The application of DSSCIM is explained in Chapter eight.

4.8 REFERENCES

1. Berliner, C. and Brimson, J.A., 1988, *Cost Management for Today's Advanced Manufacturing*, Harvard Business School Press.
2. Bhat, V.N., 1992, "Applications of the cost-value technique for the AMT acquisition", *Proceedings of International Engineering Management Conference on Managing in a Global Environment*, USA, pp. 179–83.
3. Boaden, R.J., and Dale, B.G., 1990, "Selecting systems to support computer integrated manufacturing", *Engineering Costs and Production Economics*, Vol 20 No 3, pp. 305–318.
4. Chen, I.J. and Small, M.H., 1994, "Implementing advanced manufacturing technology: an integrated planning model", *Omega International Journal of Management Science*, Vol 22 No 1, pp. 91–103.
5. Meredith, J., 1987, "New justification approaches for CIM technologies", *CIM Review*, pp. 37–42.
6. Primrose, P.L., 1991, *Investment in Manufacturing Technology*, (London: Chapman and Hall).
7. Saaty, L.T., 1980, *The Analytic Hierarchy Process: Planning, Priority Setting, Resource Allocation*, (New York: McGraw-Hill International).
8. Skinner, W., 1985, *Manufacturing the Formidable Competitive Weapon*, John Wiley and Sons.
9. Weatherall, A., 1992, *Computer Integrated Manufacturing: a total company competitive strategy*, (Great Britain: Butterman-Heinemann).

CHAPTER FIVE

Assessing the Impact of Investment in CIM

5.1 OVERVIEW

How do we assess the impact of investing in an AMT in one functional unit over others, when technologies considered for investment are to be integrated, consist of many intangible benefits, improves with the learning-curve effect and is expected to give benefits to other functional units due to synergy as well? This chapter presents the answers for the above question as the second part of the systematic integrated approach in relating strategic consideration of the enterprise for investing and evaluating the investment. The first part of the integrated approach is discussed in Chapter four.

This chapter opens with a discussion on cost-impact evaluation on functional units and then explains a financial methodology utilised within the integrated justification and optimisation approach of CIM.

5.2 COST-IMPACT EVALUATION IN FUNCTIONAL UNITS

Preliminary estimation of the effect of intangibles on an investment decision will provide adequate knowledge about the investment to the decision-maker. This procedure of estimation will help identify the pitfalls and consequences of the investment decision before actually proceed to a financial analysis. The impact of AMT is not limited to the functional unit, where the AMT has been installed. The cost-impact evaluation in functional units is handled as the fourth stage in the proposed approach. At this stage of evaluation, cost-impacts among various functional units can be estimated. However, if preferred, this stage can be avoided by the decision-makers. By analysing the cost-impact of selecting an AMT, and treating the company as a single entity, the decision-makers are in a better position to investigate the effect of investment on the company by the strategic attributes.

By utilising a functional inter-relationship matrix and an impact evaluation analysis, the cost-impact of investing in subsystems on each functional unit and on the company can be generated while considering the company as a single entity. The methodology provides two options such as detailed and system-wide analysis to evaluate the cost-impacts on subsystems (see Figure 4.1, page 55).

For the option of detailed analysis, out of the selected attributes, the attributes which are closely linked to the selected subsystem, are analysed as shown in the Venn diagram of Figure 5.1 (page 71). For the system-wide option, the effect on the functional unit as a whole is estimated. These tentative evaluations need to be carried out only if a necessity exits. Otherwise, we can move towards the stage of evaluating financial measures.

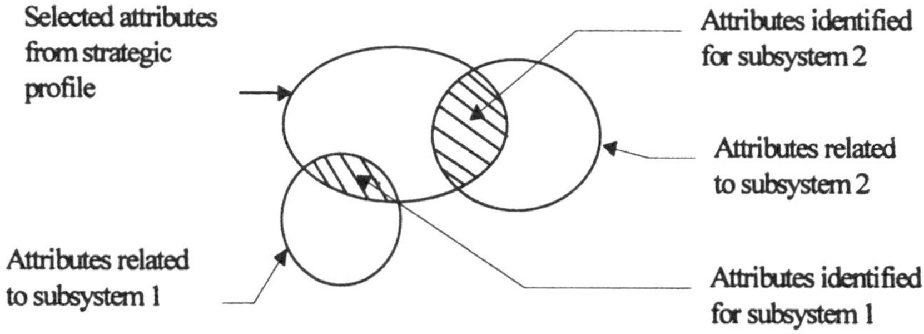

Figure 5.1 Venn diagram for the display of related attributes

5.2.1 Theory on Evaluating Cost-Impact

As we have seen earlier, investments in AMTs are interrelated. The equations and the fundamentals described in this section will assist decision-makers and manufacturing managers to realise the impacts on the other systems and functional units due to the implementation of an AMT in one functional unit. The relationships identifying cost-impacts can be expressed as:

Cost-impact by an AMT	=	Cost-impact in prime functional unit by that AMT	+	Cost-impact in other functional units by synergy

and

Cost-impact in a functional unit	=	Cost-impact by the technology invested in that functional unit	+	Synergy effect by other technologies on other functional units

For instance, if CAD is considered as a prime technology in Design and Engineering (D & E) functional unit, the cost-impact by CAD and the cost-impact on D & E are formulated as:

Cost-impact by CAD	=	Cost-impact in D & E functional unit by CAD	+	Cost-impact in other functional units by CAD
Cost-impact in D & E functional unit	=	Cost-impact by CAD in D & E functional unit	+	Cost-impact by other technologies (CAM, CAPP, CAQC, etc.) in D & E functional unit

5.2.1.1 Preliminary cost-impact analysis

Tentative cost-impact evaluation relies on the preliminarily estimation of the effect of intangibles provided by the decision-maker. It can be calculated as follows:

Let

a_{ij} = Functional inter-relationship ratio

X = Total operating cost of the company

Y_j % = Cost share factor of the department (j)

Z_k % = Cost-impact on each attribute (k)

Z_{Tij} = Cost-impact on a technology Ti at the prime functional unit (j*)

= $\sum_k Z_k$, (Where j* denotes prime functional unit)

The prime functional unit for a technology is defined as the functional unit in which the technology is assigned to work primarily. By utilising the above notations, the derivations for the cost-impacts can be expressed as:

$$\text{Cost-impact by technology } T_i = X * Z_{Tij} *\left(\sum_{j=1}^{n} a_{ij} * Y_j\right) \quad (5.1)$$

(since $a_{ij} = 1$ for prime technology at the prime functional unit)
and,

$$\text{Cost-impact in a functional unit } FU_j = X * Y_j * \left(\sum_{i=1}^{m} a_{ij} * Z_{T_{ij}}\right) \quad (5.2)$$

Then, the overall cost-impact within the company as a single unit is:

$$\sum_{i=1}^{m}\{X * Z_{T_{ij}} * (\sum_{j=1}^{n} a_{ij} * Y_j)\} = \{X * Y_j * (\sum_{i=1}^{m} a_{ij} * Z_{T_{ij}})\} \quad (5.3)$$

The percentages of cost-impacts can be calculated as follows:

The percentage of the cost-impact by technology Ti

$$= \frac{\text{Overall cost - impact by the technology } T_i}{\text{Overall cost - impact within the company}} \% \quad (5.4)$$

The percentage of cost-impact in a functional unit FU j

$$= \frac{\text{Cost - impact in a functional unit } FU_j * 100}{\text{Overall cost - impact within the company}} \% \quad (5.5)$$

The percentage of overall cost-impact within the company

$$= \frac{\text{Overall cost - impact within the company} * 100}{X} \% \quad (5.6)$$

Calculation of direct and indirect cost attributes can be performed to a reasonable degree of accuracy, by utilising the standard relationships available in manufacturing textbooks. The impact of subsystems can be evaluated again on completion of the financial analysis stage as well. For this computation, the data from the financial analysis stage are to be utilised along with the functional interrelational matrix. At this stage, the decision-makers will have a clear picture of the investment associated with the implementation of AMTs.

5.2.2.2 Definitive cost-impact analysis

Financial analysis provides a mechanism to obtain the total operating savings and cost for every investment considered in the analysis. Once the financial indices are calculated for the selected technologies, calculation of cost-impacts will provide an realistic indication of the impact of the technologies on the other functional units, other technologies and on the company as a single entity. The cost-impacts can be computed according to the following relationships:
Let
TO_i = Total operating savings for technology (i)
TC = Total operating cost of the company in all functional units
Then,

$$\text{Cost-impact by technology } T_i = TO_i * (\sum_{j=1}^{n} a_{ij}) \qquad (5.7)$$

The percentage of cost-impact by a technology T_i

$$= \frac{\text{Cost - impact by technology } T_i * 100}{\text{Sum of total cost in all functional units}} \% \qquad (5.8)$$

similarly,

$$\text{Cost-impact in a functional unit } FU_j = \sum_{i=1}^{m} (a_{ij} * TO_i) \qquad (5.9)$$

The percentage of cost-impact in a functional unit FU_j

$$= \frac{\text{Cost - impact in a functional unit } FU_j * 100}{\text{Sum of total cost in all functional units}} \% \qquad (5.10)$$

$$\text{The overall cost-impact in the company } = \sum_{i=1}^{m} TO_i \qquad (5.11)$$

$$\text{Then the percentage of overall impact } = \frac{\sum_{i=1}^{m} \{TO_i * (\sum_{j=1}^{n} a_{ij})\}}{TC} \% \qquad (5.12)$$

5.3 FINANCIAL ANALYSIS OF THE INVESTMENT

Existing cost accounting systems and cost management practices do not adequately support the objectives of manufacturing automation and do not have a facility to evaluate intangibles, which are inherent in CIM systems. Due to this impasse, manufacturing managers and decision-makers do not have a clear concept in justifying the potential intangible and indirect costs/benefits. When the decision-makers find quantifying these intangible benefits as a difficult process they tend to either omit or ignore these benefits completely from the financial appraisal. The omission of these benefits may lead to the rejection of investment in AMTs, unless companies modernise their cost management practises or invest in AMTs for the strategic purpose disregarding the results of the financial evaluation.

As AMTs make the companies move away from labour intensive production to less labour involvement, incorporation of strategic factors is essential to justify the investment. We have briefly looked at some of the traditional decision-making techniques in Chapter three. However, the shortcomings explained in Chapter three, Section 3.4, necessitates a new accounting system that can support management decisions by calculating values for the real costs and values of lead-time, space utilisation, inventory and quality. Any investment in new technologies must be aimed at achieving financial objectives rather than investing for the sake of technological change. As explained in Chapter four (Section 4.3), the evaluation of financial measure is handled as the fifth stage in the procedure and it can be evaluated by either of the two options suggested in the procedure of the DSSCIM (see Figure 4.1, page 55). The most preferred option incorporates the Cost-Value technique in the evaluation of the intangibles. The other is a standard approach preferred by managers, where intangibles are disregarded.

5.3.1 A Methodology to Estimate Financial Benefits of Attributes

Disintegration and re-quantification of the attributes can be utilised to estimate many intangible benefits resulting from the AMTs. Disintegration defines an approach where the intangible attributes are decomposed into many tangible and quantifiable attributes. This approach was earlier suggested by Primrose and Leonard (1986). A typical example of disintegrating the intangibles had been explained by them, for the attribute of "Improved product quality". They had pointed out this attribute could be represented by reductions in scrap, rework, disruption of production, warranty payments and increased sales of better quality products. On the other hand, re-quantification explains a technique where an intangible attribute is looked at in a different perspective for defining it. This technique was initially suggested by Catton (1986). His alternative method of quantifying intangible attribute is by asking the question of "so what?" when confronted with an intangible attribute. By this questioning some of the intangibles could be converted into tangible attributes.

However, some intangible attributes cannot be disintegrated or re-quantified. On the other hand, learning from the leaders who have already implemented AMTs is a great help for manufacturing decision-makers, who are striving to justify an investment, on estimating benefits and costs. A database, which is established from the available literature, can be utilised for the estimation of savings in this approach. Hence, in this book, both methods have been combined to provide quantification of some of the tangible and intangible attributes. Although, for some tangible attributes, mathematical derivations can be applied directly to estimate the benefits, the combination of disintegration and estimation will provide acceptable results as well. In addition, even in mathematical equations, the expected benefits need to be estimated.

The following illustrates the method of estimation utilising the database for the attributes considered in the analysis. In addition, wherever possible, mathematical derivation is also given to the attributes.

Attribute: Minimising direct labour

Over the years, many manufacturing managers have attempted to justify investments in AMTs solely based on reduction in labour costs. The labour costs could consist of overtime benefits, which usually incur higher rates than the normal regular rates and benefits paid during the leave entitlement. The labour costs are generally dependent on the functional units which adopt a new technology. However, the introduction of an AMT will reorient many employees after retraining into different functional units, and will raise their useful level to multi-skilled and multi-functional oriented. Conversely, replacing the employees by any new technology is a controversial issue in management concepts. The reduction in the number of workers by the implementation of an AMT can result in additional costs such as the cost of retraining transferred workers to other functional units, termination benefits paid to the dismissed workers and legal and managerial costs incurred from union disputes. Hence, it is advised that estimation of the benefits should be conducted carefully and cautiously considering the above facts. The saving due to the minimisation of direct labour is calculated in the methodology as follows:

Let

Existing labour units = ELU/number

Wages/labour = \$WpL

Hence

$$\text{Savings} = \text{ELU} * \text{WpL} * \text{percentage savings from the database}/100 \qquad (5.13)$$

Using mathematical derivations:

Let

Categories of labour associated with AMT = I (I = 1 to n)

Existing labour units in each category for the AMT = L_i/number

Wages for labour in category I = \$$W_i$

(including overtime, and government and administrative charges)

Expected reduction in each category I = X_i/number

Hence

$$\text{Savings} = \sum_{i=1}^{n}(X_i * L_i * W_i) \qquad (5.14)$$

Attribute: Reducing material costs

Material costs in a manufacturing environment consist of the actual raw material utilised in the products, the wastage and tooling factor allowances considered for the production process, and the rejected parts to manufacture certain number of accepted parts after inspection. Since rework costs and scrap costs are considered separately in the analysis suggested in this methodology, the definition of material cost is limited to the actual raw material utilised in the products and the wastage and tooling factor allowances considered for the production process. Utilising an AMT such as CAM will lead to reduction of wastage and tooling factor allowances. This reduction will result in savings towards the purchase of actual raw materials. The saving due to reduction in material is calculated in the methodology as explained.

Let
Existing material cost = \$EMC
Hence,

$$\text{Savings} = \text{EMC} * \text{percentage savings from the database}/100 \tag{5.15}$$

Using mathematical derivations:
Categories of products produced at AMT = I (I = 1 to n)
Number of units produced in each category /year = U_i /number
Raw material cost for category I = \$$C_i$ /unit
Expected material cost reduction in category I (due to tooling and waste) = X_i /unit
Hence

$$\text{Savings} = \sum_{i=1}^{n}(X_i * U_i * C_i) \tag{5.16}$$

Attribute: Reducing material handling costs

Material handling costs involve mainly direct and indirect labour costs related to movement of work parts from one place to another. Introduction of AMTs eliminates tedious repetitive tasks and provides the capability to deliver different parts to different manufacturing systems according to a planned routing. In addition, the reduced product losses from handling damage, location errors and improved security in material handling produces beneficial financial savings for a company. Especially, robots and AS/RS reduce the raw goods inventory in a manufacturing environment. Hence, the saving due to reduction in material handling is calculated in the methodology as follows:

Let
Existing material handling cost = \$EMHC
Hence

$$\text{Savings} = \text{EMHC} * \text{percentage savings from the database}/100 \tag{5.17}$$

Attribute: Reducing tooling/fixture costs

The cost of tooling and fixtures are based on the dedicated tooling and fixture arrangements necessary in an AMT environment. In addition, it includes the worn-out and replacement tools due to tool failure. It has been found that adopting AMTs significantly reduces this cost, since an AMT environment requires less dedicated fixtures and tools than the conventional machines. The saving due to reduction in number of tooling/fixtures is calculated as follows:

Let
Existing tooling cost = \$ETC
Existing fixture cost = \$EFC
Hence

$$\text{Savings} = (\text{ETC} + \text{EFC}) * \text{percentage savings from the database}/100 \tag{5.18}$$

Using mathematical derivations:
Existing tooling and fixture cost per unit = T
Number of units = n

$$\text{Hence Savings} = T * n - \text{Expected costs in AMT environment} \tag{5.19}$$

Attribute: Reducing scrap costs
Scrap is an inevitable product in a manufacturing environment. However, automation, especially in AMTs, reduce the scrap related costs considerably. Hence, the saving due to reduction of scrap costs is calculated in the methodology as follows:
Let
Existing scrap cost = $ESC
Hence,

Savings = ESC * percentage savings from the database/100 (5.20)

Using mathematical derivations:

Savings = Existing scrap costs – Expected costs in AMT Environment (5.21)

Attribute: Reducing maintenance costs
The maintenance cost in an AMT environment involves the software and hardware related costs. Automated systems have a higher tendency to be utilised than their conventional counterparts. Hence, the maintenance costs may increase rather than decrease. However, a CNC or a CAM system can be equivalent to many dedicated manufacturing machines, then the overall maintenance cost will be reduced by the investment in AMTs. However, due to the complexities involved in AMTs, the new equipment may experience higher maintenance costs per service than the traditional equipment. Therefore, caution should be exercised when evaluating costs. The saving due to reduction in maintenance is calculated as follows:
Let
Existing maintenance cost = $EMC
Hence

Savings = EMC * percentage savings from the database/100 (5.22)

Using mathematical derivations:
Existing maintenance cost of a dedicated machine = C
Number of dedicated machines = n

Savings = C*n – Expected costs in AMT environment (5.23)

Attribute: Reducing production planning and control costs
Costs related to production planning control mainly arise from indirect labour involved in producing appropriate production and process plans to manufacture a component. The use of CAPP will significantly reduce the above cost. Hence, saving due to reduction in production planning control is calculated as follows:
Let
Existing production planning and control cost = $EPPC
Then

Savings = EPPC * percentage savings from the database/100 (5.24)

Using mathematical derivations:
Existing labour (direct and indirect) costs on production and planning = L
Number of labour involved = n

Savings = L*n – Expected costs in AMT environment (5.25)

Attribute: Reducing floor space costs

Floor space for certain AMTs such as CAQC tends to be expensive due to the requirement of the environmentally controlled space with shock absorbing facilities. On the other hand, robots can work in hazardous environments and in very limited space. AS/RS will dramatically reduce the floor space required for inventory and working. Generally, many of the AMTs require less floor space. Hence, the total cost of floor space for the AMTs will be less than investing in conventional machines for the equal productivity output. The saving due to reduction in floor space is calculated as follows:

Let

Existing floor space cost = $EFSC

Then

Savings = EFSC * percentage savings from the database/100 (5.26)

Using mathematical derivations:

Savings = Existing floor costs – Expected costs in AMT environment (5.27)

Attribute: Reducing inspection costs

Inspection adds a significant amount of indirect labour into a manufacturing procedure. Utilisation of AMTs will substantially reduce the inspection costs associated with manufacturing a product. Especially, the use of CAQC reduces the cost and time involved in inspecting a produced component for quality standards. The use of automated inspection facilities provides inspection of all production output rather than sampling inspection procedures resorted to by conventional methods. The saving due to reduction in inspection is calculated as follows:

Let

Existing inspection costs = $EIC

Hence

Savings = EIC * percentage savings from the database /100 (5.28)

Using mathematical derivations:

Number of inspectors for quality testing = n

Cost per inspectors = C

Cost of existing testing equipment = E

Hence

Existing inspection costs = E+nC

Savings = E + nC – Expected costs in AMT environment (5.29)

Attribute: Increasing manufacturing productivity

Manufacturing productivity increases as a result of increase in throughput and labour productivity, and reduction in product cost, cycle time, downtime, and setup time. In reality, productivity improves due to the fact that less manufacturing time is taken to produce the same amount of output. Hence, considering the time as a parameter in the computations may lead to a double counting error, which is an accumulated error with respect to counting an attribute more than once in a computation.

Productivity can also be defined as total output divided by total input. However, in this book, the definition of productivity is limited to the concept of the capacity to produce more items. In addition, if increased throughput and the reduced product cost are considered as factors in determining the savings, the computation will also be affected by the double counting error. The reduction in product cost occurs due to the increased ability to produce more by way of the reduction in setup time and idle time.

Let

Existing throughput/year = ETP

Unit production cost per unit = UPC

Hence

Savings = ETP * UPC * percentage savings from the database/100 (5.30)

Attribute: Improving machine utilisation

Improvement in machine utilisation occurs due to the reduction in idle time of the manufacturing process. Usually a 50-week cycle for a year is considered for a commercial enterprise, when analysing financial impacts. Hence, the saving due to improvement in machine utilisation is calculated as follows:

Let

Existing idle time/week = EIT /hr

Operating cost/week = $OC

Number of hours used/week = NU

Hence

Savings =

EIT * OC * NU * 50 * percentage savings from the database/100 (5.31)

Attribute: Reducing work-in-progress costs

Work-in-progress (WIP) represents products either being processed or waiting to be processed at the production line. Application of AMTs extensively reduces the WIP inventory in a company, thus leading to savings in WIP-related costs. The saving due to reduction in WIP cost is calculated as follows:

Let

Existing WIP costs = $EWC

Hence

Savings = EWC * percentage savings from the database/100 (5.32)

Attribute: Reducing rework costs

Rework is the result of a fault in assembly or any manufacturing error in production process that can be rectified. The use of AMTs will greatly reduce the rework related costs in a manufacturing environment. Thus, the saving due to reduction in rework is calculated as follows:

Let

Existing rework costs = $ERC

Hence

Savings = ERC * percentage savings from the database/100 (5.33)

Attribute: Reducing lead-time

Lead-time is considered as the total time required to produce an item from the order received. In today's competitive world, reduction in lead-time will enable manufacturing companies to market their products quickly and competitively, thus time, as a strategic parameter, plays a major role in capturing more market share and profit. By understanding the time-based competitiveness, time can be managed in accordance with the corporate objective. The lead-time for a product may consist of:

1. waiting time to get the raw materials from suppliers;
2. manufacturing lead-time for the product;
3. queue time;
4. transportation time;
5. inspection time; and
6. packaging time.

However, manufacturing lead-time can be limited to the total time required to process a given product in a plant. The reduction of lead-time will enable production of more items at the same time. Manufacturing lead-time can be expressed by Equation 5.34 (Groover, 1987).

$$\text{Manufacturing lead-time} = \sum_{i=1}^{n} (Tsu_i + Q * Top_i + Tnop_i) \qquad (5.34)$$

Where
I = The operation sequence (I = 1 to n)
Tsu_i = Setup time for the equipment
Q = Quantity of items produced
Top_i = Operating time for each item
$Tnop_i$ = Non-operating time for the equipment

However, in the methodology proposed in this book, the saving due to reduction in lead-time is calculated as follows:
Let
Existing lead-time of a product = ELT/hour
Operating cost/hour = $OC
Number of items produced/week = NIW
Hence
Savings =
ELT * OC * NIW * 50 * percentage savings from the database/100 (5.35)

Attribute: Reducing paperwork costs
In a traditional manufacturing work environment, much of the data and information is transferred via printed documents and diagrams. In contrast, the use of AMTs will reduce the amount of paperwork involved in producing these documents as a medium for information, since AMTs will reduce the time and labour involved in duplicating and transferring the information. Most significantly, an integrated enterprise will have the data sharing facilities, thus eliminating much of the paperwork in a manufacturing environment. Thus, the saving due to reduction in paperwork is calculated as follows:
Let
Existing paperwork costs = $EPC
Hence, Savings = EPC * percentage savings from the database/100 (5.36)

Attribute: Improving quality
Quality improves visual and functional aspects of the product. A product that does not meet the agreeable quality of the customer will incur additional cost called quality cost. Quality cost can be considered in the following categories (Golomski, 1990):

1. External failure costs. This includes all the costs associated with failures outside the plant. They include product return costs, rework and scrap costs of returned products, administration costs of customer complaints, indirect labour costs involved in resolving customer complaints, distribution and transportation losses;
2. Internal failure costs. These are the costs encountered inside the plant such as scrap, rework, repair and associated costs due to those aspects;
3. Appraisal costs. They include cost of performing quality inspection internally and externally; and
4. Prevention costs. This category involves quality training and educational programs to improve quality of the product.

Advanced manufacturing technologies improve the quality of a product. Aspect of quality in AMTs has generated much interest. Improved quality will result in some or all of the following:

1. reduction in inspection cost;
2. reduction in inspection time;
3. reduction in rework cost;
4. reduction in scrap and rejection cost;
5. reduction in cost of discounts offered on substandard products;
6. reduction in warranty payments and service costs; and
7. increase in revenue from additional business gained.

In the proposed methodology, to provide simplicity in the analysis of the attribute "improving quality", the attribute is limited to the consideration of reductions in the areas of scrap, rework, warranty payments and inspection cost. Quality is not considered separately as an individual attribute.

5.3.2 Learning Curve Effect

The concept behind the learning curve was first published by T.P Wright in 1936 in the article "Factors Affecting the Cost of Airplanes" (cited in Stewart, 1991). Ever since, this concept has been widely accepted. The learning curve is an explanation for the well observed phenomenon of increased worker efficiency and improvements in repetitively performed tasks. The basic concept of the learning curve is as the units of production increases, cost and production time decreases. Similarly, as the experience with the AMT increases, the benefits gained from them will increase over time. Hence, a simple exponential equation was added in the proposed methodology to represent this effect over the years. Once the benefits due to the attributes are calculated for the first year, we can use the suggested mathematical equation for the learning curve to represent the values gained by these intangibles over the years within the selected study period.

$$B_n = B_1 * \alpha^{(n-1)} \tag{5.37}$$

Where B_n = Benefits at the n th year
n = The number of years of the study period (n = 1 to 10)
α = Index for the learning curve effect

5.4 APPLICATION OF COST-VALUE TECHNIQUE FOR QUANTIFICATION OF INTANGIBLES

The intangibles are difficult to quantify and cannot be accounted for by current financial methodologies. This section presents an effective and simplified approach to solve the problem created by intangibles. This approach incorporates an improved Cost-Value technique with current financial methodologies, for the evaluation of intangibles for the justification of CIM and/or its components. The Cost-Value technique has been utilised in the past for the selection of suitable computer hardware and software among the various vendors, competing for tenders. The same methodology can be used to quantify the intangibles by assessing the value of each intangible for the company concerned with appropriate modifications. The proposed methodology effectively utilises the application of the Cost-Value technique for the justification of AMT or CIM.

As Joslin (1968) suggests, evaluating the values of the attributes before receiving the proposals from vendors will eliminate the bias in decision-making towards some vendors. Accordingly, in our methodology, we evaluate the attributes of each technology before making any purchase decision. When the value of the intangible attribute is established for an AMT, it will overcome the barrier in the justification process. It is expected, having a clear concept in justifying the intangibles related to AMTs and CIM investment, will induce the companies towards making use of the advanced technologies without any hesitation. Conducting an evaluation with the Cost-Value technique requires adopting the following two steps:

- Step 1: Identifying the technology considered for the investment.

In any justification issue it is mandatory to identify the technologies which are suitable for consideration based on strategic and tactical forecasts. In our methodology, we identified a method to select the required technology based on the strategic requirements of the company. For the selected technology considering tangible and intangible attributes, which are associated with it on the knowledge base, is a significant improvement from the Cost-Value technique originally proposed by Joslin (1968), in which the attributes of an AMT have to be categorised into desirable and mandatory features. This improvement in our methodology results in simplifying the technique into an effective tool to evaluate all the values of the attributes related to that technology disregarding the difference of an attribute being desirable or mandatory. To guide the decision-maker, the attributes associated with every selected AMT for evaluation are automatically provided by the DSSCIM, which was developed by the authors, in a similar manner to the depicted relationship in Venn diagram of Figure 5.1 (page 71). This provision enables the decision-maker to be aware of the tangible and intangible attributes associated with an investment in AMT.

- Step 2: Identifying the value of an attribute

This is performed by asking the question: What is the value of an attribute of the technology within a functional unit, where the technology is going to be implemented? The use of the Cost-Value technique requires identification of the perceived value of each intangible attribute associated with the selected technology. Generally, the value of an item is considered less than the cost of the item, when that item is competitively available. This criterion is applicable for attributes sought by a company to achieve its strategic decision. However, there are some exceptions for this criterion, where some of the general items will have more sentimental or utility value than its cost.

In a manufacturing environment, the general rule of value is less than the cost of an item is applicable. Consider a technology under evaluation that has an additional facility to count the number of items processed through it at the additional cost of $5,000. However, if there is no requirement to count the number of items passing through that machine in the manufacturing procedure, the value of having that additional feature on that machine is zero for the company. Nevertheless, if that facility can be used in future operations of the company, a moderate value such as $800 can be assigned. As Primrose and Leonard (1986) has stated many of the intangible attributes could be broken into tangible attributes and represented accordingly. These representations could be quantified within a reasonable accuracy with little difficulty. On the other hand, if an intangible attribute cannot be divided into a number of tangible attributes, the usage of Cost-Value method will be the only available option to quantify that intangible attribute.

5.5 UTILISATION OF COST-VALUE METHOD

The effective use of the Cost-Value technique requires assigning a monetary value, for each important attribute, depending on the company's requirement. During the financial analysis stage, in our methodology, it will be useful to the decision-maker to use a financial-analysis-sheets similar to that as shown in Table 5.2 (page 84), with the similar format for each subsystem. When using Cost-Value to quantify attributes directly related to the subsystem, we need to estimate both the benefits from an attribute and the value of an attribute for the company. Equations 5.13 to 5.36 can be utilised by the decision-makers to obtain the value of intangibles and benefit from them. In this approach, the perceived benefit from an attribute is incorporated with other benefits. However, the value of an attribute related to the subsystem, where the AMT has been installed, is deducted from the capital cost according to the Cost-Value method for comparison against investing in other alternative technologies. Caution should be exercised to avoid counting the benefits twice as a value and as an improved saving. In the proposed methodology, the value and benefits are considered separately, since both are different entities in any evaluation procedure.

Table 5.2 A template for financial analysis

Technology	Robotics					
Prime Functional Unit	Assembly					
Year	0	1	2	3	4	5
Cost-Value analysis incorporated						
	(in1000s)					
Investment, Capitalised						
Equipment Cost	1200.0					
Accessories & Tooling Cost	30.0					
other---	8.0					
Total Capital Inv	1238.0					
Tax Credit for the investment	10.0%					
Investment Tax Credit	123.8					
Effective Cap. Investment	1064.2					
Investment, Expended						
Engineering	35.0					
Installation	20.0					
Start-up	10.0					
Other	3.0					
Total Expensed Inv.	68.0					
Operating Savings (Costs)						
Saving Robot Value		981.90	1001.54	1021.57	1402.00	1062.84
Other Savings						
Other Costs						
Total Operating Savings	–68.0	981.90	1001.54	1021.57	1402.00	1062.84

For instance, in the case of the Cost-Value option, if the organisation is willing to spend approximately $7,000 for improving quality of the products, or it perceives the value of improved quality in the organisation as worth $7,000, then we can assign that amount of money for improving quality. In other words, this technique of estimating a value for an attribute will force us to establish a reliable monetary value for an attribute. Assigning a value is a complex and abstract method. The subjective measurement of value of an attribute may change or vary over time depending on the importance of that attribute to the company. By deducting the value of the desirable attributes from the total capital cost of the investment, the company will be in a better position to justify an investment in AMT compared to investing in alternative technologies using existing financial techniques.

Manufacturing companies invest money to make a return on the investment. Hence, all the value assignments should be further analysed to verify whether the perceived value on the strategic attributes directly or indirectly improves the financial position of the company within a reasonable timeframe. For example, improved product quality may result in an increase in sales of the items, and leads to capturing more market share, or the product with improved quality can be offered to the customers at a higher price, whereby increasing the profit margin.

Manufacturing managers determine the after-tax-cash-flow analysis using financial techniques such as NPV, IRR and PB period of the investment. These preferred financial results are calculated, for the benefit of the decision-makers. In addition, cash flow is calculated after considering depreciation of the assets using one of the methods among depreciation methods such as straight-line depreciation, sum-of-years depreciation and fixed declining balance method. A typical template suggested for the methodology and which is utilised in the DSSCIM for the computation of the financial results is shown in Table 5.3 (page 86). Using the templates for the financial analysis (Table 5.2 and Table 5.3), the following calculations are computed by the DSSCIM for each technology selected by the decision-maker. However, a decision-maker who does not want rely on the DSSCIM can perform this analysis manually according to Equations 5.38 to 5.43.

Effective capital investment =
Total capital investment – Investment tax credit – Values of attributes (5.38)

Total operating savings =
Operating savings (costs) by tangible attributes + Savings by intangible attributes – Total expended investment (5.39)

Net before tax savings = Total operating savings – Depreciation (5.40)
Net after-tax savings = Net before tax savings *(1 – Tax rate) (5.41)

Net after-tax cash savings = Net after tax savings + Depreciation (5.42)

Net after-tax cash flow =
Net after tax cash savings – Effective capital investment + Investment tax credit (5.43)

Although the templates in Table 5.2 (page 84) and Table 5.3 (page 86) show only the first five years in the analysis period, the decision-makers can extend or shorten the analysis period depending on their requirement. The category "other-costs" in Table 5.2, should include overhead costs including the cost of maintenance, and power/energy related costs to reflect the changing cost mixture of the automated investment. When the Cost-Value option is not used, the effective capital investment equals the total capital investment, since attributes are not assigned with any monetary values. "Effective capital investment" is used as an index to compare the capital costs related to investing in conventional technologies and to calculate NPV, IRR and PB period. However, "Total capital investment" is utilised to perform depreciation related calculations, since the cost of acquiring an asset and the depreciation of it will not be changed by the value of attributes the AMT offers to the company.

Table 5.3 A template for computation of financial results

Depreciation	0.00	151.00	151.00	151.00	151.00	151.00
Net Before Tax savings	-68.00	830.90	850.54	870.57	891.00	911.84
Tax Rate %		0.15	0.15	0.15	0.15	0.15
Net After-Tax Savings	-68.00	706.27	722.96	739.98	757.35	775.06
Net After-Tax Cash Savings	-68.00	857.27	873.96	890.98	908.35	926.06
Net After-Tax Cash flow	-1008.40	857.27	873.96	890.98	908.35	926.06
Discount Factor %	20.00	20.00				
NPV @ d.f% at end of 10 years	2451.79					
IRR %	86.30					
discounted P.V		714.39	606.92	515.61	438.05	372.16
Pay Back Period	1.50					
Total Cost = Total Capital Inv. + Total Expensed Inv.	1306.00					

5.6 CONCLUDING REMARKS

Advanced manufacturing technologies have an inherent difficulty in justification due to the many intangible attributes associated with them. Due to this difficulty, managers tend to avoid the intangible attributes within the justification process. This avoidance imposes a severe limitation on the justification procedure. Any justification should be made on complete information without ignoring any features that are difficult to incorporate into the procedure. The Cost-Value technique has a disadvantage of being a subjective estimation procedure. However, the subjective estimation of intangibles is better than ignoring intangibles, since it enables the final decision to be closer to being right than being wrong. It has been found, although the Cost-Value technique is a subjective evaluation procedure, it will enable the decision-makers to analyse investment decisions with the current financial evaluation procedures more effectively. This is significant because any future investment analysis cannot avoid a subjective assessment, even if it is for a single equipment replacement justification. Further, the consideration of intangible attributes within the justification framework enables a decision-maker to derive decisions based on complete information. Incorporating the values of qualitative attributes into the justification procedure will definitely justify the costs involved in investing in AMTs. However, the subjectiveness in reaching a decision is very much reduced by the mechanism of the Cost-Value technique when it is used in conjunction with disintegration of attributes.

To reduce the subjectiveness of the Cost-Value technique, further research must be conducted to redefine the intangibles associated with an investment into tangible attributes. Meanwhile, few companies have become leaders in the field of CIM. Learning from them would be a valuable experience for manufacturing managers interested in implementing CIM in their companies. Hence, establishing a comprehensive database that provides benefits accrued for each AMT at the leading companies will effectively help reduce the subjectiveness associated with estimating the tangible and intangible attributes.

This chapter presents an approach to evaluate the company-wide impact of investing an AMT in one functional unit and a methodology to estimate financial benefits of attributes within the integrated approach to CIM justification and optimisation. This is the second part of the integrated approach, which is the continuation from the previous chapter. The DSSCIM developed by the authors incorporates these methodologies to evaluates the impact of the investment in various functional units and quantify the intangible benefits while pursuing the strategic objectives of the company. The next chapter (Chapter six) describes the last stage of the integrated justification and optimisation approach as the third part in the sequence.

5.7 REFERENCES

1. Catton, P.G., 1986, "Advanced manufacturing technology - evaluating the intangible benefits", *Proceedings of Planning for Automated Manufacture*, London, pp. 17–22.
2. Golomski, W.A.J., 1990, "Accounting aspects of automated inspection systems", edited by, Parsaei, H.R., Ward, T.L. and Karwowski, W., *Justification Methods for Computer Integrated Manufacturing Systems: Planning, Design, Justification and Costing*, (Amsterdam: Elsevier Science), pp. 242–249.
3. Groover, M.P., 1987, *Automation, Production Systems, and Computer Integrated Manufacturing*, (New Jersey: Prentice-Hall Int. editions).
4. Joslin, E.O., 1968, *Computer Selection*, (USA: Addison-Wesley Publishing).
5. Primrose, P.L. and Leonard, R., 1986, "The financial evaluation and economic application of advanced manufacturing technology", *Proceedings of Planning for Automated Manufacture*, pp. 49–54.
6. Stewart, R.D., 1991, *Cost Estimating*, (USA: Wiley-Interscience Publications).

CHAPTER SIX

CIM Optimisation

6.1 OVERVIEW

How do we make an important decision? We must know what are the requirements we want to achieve and what are the limitations we face to reach our goals. Similarly, decision-making in an investment appraisal prerequisites the goals of the organisation to be determined in advance. As we have realised in previous chapters many commercial organisations generally determine their analysis only on the profit maximisation. In contrast, implementation of AMTs can offer many benefits for a company to achieve its business objectives in addition to increasing its revenue. In today's context the business objectives are diverse and require a multicriteria decision-making approach to determine the optimal resource allocation for capital investments.

Optimisation is utilised in all operations where the limited resources have to be efficiently controlled and managed. Optimisation is a mathematical and numerical approach to identify the best procedure among existing and competing alternatives within the identified constraints. In the previous chapter we were looking for an optimisation approach, which we could use to integrate with the justification approach for CIM. Goal programming (GP), one of the multicriteria optimisation approach, provides an opportunity to reach an optimal decision in a multicriteria environment. This chapter briefly introduces an optimal resource allocation methodology based on a GP algorithm, which could satisfy the strategic objectives of the company. The suggested methodology is incorporated into the integrated justification and optimisation approach proposed by the authors to optimise the results as the third part of the procedure.

6.2 NEED FOR CIM OPTIMISATION

There are many optimisation problems present in most fields of science, technology, engineering, economics and management. In industry, the necessity for optimisation arises due to the diversity of the goals an organisation would have to consider for achieving success. The goals of the organisation are specific values the organisation holds as important to them. In addition, the goals are fully identifiable and transferable with the needs and desires of the decision-makers. However, these goals need to be defined in terms of either attributes or objectives with an established priority to fully realise them in a realistic manner. The major objectives which motivate the companies for the adoption of AMTs have been listed in Chapter one (Section 1.2.3). In addition, an enterprise may consist of diverse goals to be successful. An organisation which desires to achieve success could possibly consider the following goals:

1. Maximising profits;
2. Maximising sales and market share;
3. Achieving satisfactory level of profits;
4. Optimal use of available facilities;
5. Improving productivity;
6. Achieving target market share,
7. Utilising latest technologies to improve company's image;
8. Reducing human error;
9. Maximising employees' benefits and salaries;
10. Maintaining minimum level of employee turnover; and
11. Good working environment for employees.

As we are intrinsically aware, the absence of goals will hinder the decision-making exercise in reaching an optimal decision and will not help identify the best combination of the alternatives under consideration. On the other hand, having precise levels of goals or vaguely defining the goals will provide an opportunity to measure the desired levels of attainment in any expected improvement by an investment.

Traditionally, manufacturing investments were controlled on the basis of the cost minimisation within the available resources or on the premise of profit maximisation by enhanced production facilities. In addition, manufacturing decision-makers always resort to local optimisation instead of global optimisation. These features do not provide the effective approach towards an optimised CIM system for the enterprise.

However, an optimal allocation of the system resources in CIM and integrated AMTs is a new feature which has to be considered in order to effectively utilise the resources and to improve the system performance. As a first step towards optimised CIM, an optimal combination of subsystems should be identified to form an optimal CIM system. In addition, the investment decisions for the integrated systems must be made for the optimal benefit at the organisational level instead of at individual departments.

The use of optimisation in investment appraisal provides a mechanism to efficiently allocate the limited resources to achieve the required objectives of an investment. The ability to allocate resources in each function of the business, better than the best competitor, will lead in attaining the cost leadership and the position of strong competitor to be successful in the market (Hess, 1992).

Generally, linear programming (LP) models are utilised when the single most important objective of the enterprise becomes maximising profit or minimising the cost. In today's complex and competitive environment, the goals of the enterprise cannot be reduced to a single objective. Manufacturing investments need to be controlled to achieve the optimum objectives of an investment for the whole company while satisfying the conflicting objectives existing among various units of that company. Therefore, the investments should not be limited to achieving the sub-optimum among the few units of a company. The prime objective for many commercial companies would be cost minimisation within the available resources and achieving the profit maximisation by improved sales, due to enhanced production facilities. In addition, current market conditions require additional objectives including improved quality of the products and reduced lead-time. These objectives also have to be adequately addressed to determine the level of investment. Striving towards attaining these challenging and multiple objectives, which are often conflicting, within the strategic framework to determine the level of investment, is therefore a worthwhile procedure and to be investigated thoroughly.

6.3 PITFALLS OF SUB-OPTIMISATION AND ISLANDS OF AUTOMATION

Whenever, we encounter an optimisation problem, conventionally we concentrate on finding an optimum solution within the problem area and the constraints of the problem. The conventional approach never allows us to look beyond the problem area. The conventional approach of individual optimisation on the islands of automatons or on a single functional area often lead to a local optimum within the manufacturing cell or facility. However, this approach satisfies the departmental managers, since they can prove to the senior management that their functional unit is working under an optimal condition and they can expect commendations and promotions based on their achievements. These optimisation approaches in individual functional units never allow us to achieve the global optimum for the company, since we always look for the compromises we can make on other areas to improve the area under consideration. However, it is observed in many cases that this type of focus on individual facilities, disregarding the holistic view, often leads to incorrect results.

Further, the impact of CIM on the manufacturing environment will have far reaching effects in future. Unless, we learn to make the best use of the capabilities of AMT/CIM, help others develop their full potential on using these technologies in global manner, enrich their knowledge in implementing AMT/CIM without hesitation and manage the human element in a productive manner, the chances of establishing the manufacturing industry competitively in a rapidly changing environment and economic, technological and social constraints will be limited. Currently available justification and optimisation approaches, which could not incorporate all the factors related to an AMT, is a significant barrier in utilisation of an AMT and CIM. Today, the challenge faced by manufacturing managers is to find an appropriate approach to consider all the multiple criteria related to AMT/CIM justification.

In the past selection and justification of each technology has been individually handled by the relevant department. Hence, each individual functional unit has various conflicting goals. Even today, traditional managers handle these issues separately. Planning for investment in the AMT environment is one of the most difficult tasks of a manufacturing manager. For instance, a marketing unit needs a product that can be marketed easily at a competitive price. The concerns of a manufacturing unit are efficient production and design. A finance unit considers the cost of these investments and running the enterprise in a profitable manner. A research and development unit is concerned with introducing new and innovative products. To accommodate these conflicting goals and to provide a balanced approach to the company, a multicriteria approach, which incorporates global optimisation, is indispensable to a decision-maker.

6.4 IDENTIFYING STRATEGIC FOCUS FOR GLOBAL OPTIMISATION

World competition has forced industries to find ways and means to produce lower priced, higher quality products in shorter lead times. Hence implementing an integrated manufacturing strategy requires correct forward thinking and considerable courage to face the risks associated with its implementation.

One of the guiding principals of investment management is that sound investment decisions should be evaluated on long-term strategic plans and short-term operational goals (Berliner and Brimson, 1988). Adopting an AMT is a strategic decision, hence the implications of such action must be clearly considered before proceeding further. Therefore, before any serious optimisation approaches are used to identify the appropriate system components of CIM, the strategic focus of the company should be identified and if it is not already available, necessary steps should be taken to develop the overall strategic plan of the company.

6.5 MATHEMATICAL MODELS FOR CIM OPTIMISATION

This section gives a brief introduction to the reader on previous approaches of optimisation made for CIM or FMS by other researchers, but does not try to introduce the theories of optimisation, which can be studied from various books dealing with optimisation. Approaches for optimal allocation of resources in manufacturing gained importance in late 1980s and throughout 1990s. LP, Integer programming, Dynamic programming, Markov models, Stochastic programming and Goal programming (GP) are some of the optimisation techniques that have been cited in the literature related to FMS and CIM.

Linear programming is utilised in applications where the decision problem can be reduced to rely on a single parameter. Traditionally, LP is utilised in solving the problems in investment analysis by only considering either the budgetary limitation or the expected return. A Leontief input-output model coupled with LP and AHP was utilised by Chandra and Schall (1988) to determine the economic justification of FMS. Tie (1992) proposed an LP procedure for optimising the total cost of investment in AMT within the expected net present value. Later Kuei *et al.* (1994) formulated an LP model which relies on AHP for the selection of AMT among technologies within the available budget. Integer programming is an extension of LP, when the required decision variables are to be integers. This special requirement to formulate the decision problem occurs when the LP is concerned with variables such as people, number of machines and number of production units. In the event, when only some of the variables are required to have integer values, then the LP model is referred to a mixed integer programming (MIP). An MIP model which allows a decision-maker to purchase current best practise technology in the form of indivisible transfer lines was presented by Burstein (1986). This MIP model requires the maximisation of the discounted cash flow subject to the capacity constraints.

Dynamic programming provides a mathematical procedure for determining the optimal combinations of decisions for a sequence of interrelated decisions. A stochastic dynamic programming model that derives the value flexibility depending on the ability of an FMS to respond to changing situations was outlined by Kulatilaka (1988). This model merely provides a justification approach for FMS, utilising flexibility as a criterion. Stochastic programming is a variation of LP, where the coefficients in the LP models are not constants but random variables with a probabilistic feature. One special type of discrete time stochastic process is called Markov chain. In Markov chains the probabilities involving future states of process only depend on the present state of the process and are independent on the states the chain passed through to reach the present state. A Markov model which specifies the optimal level of capacity to convert the batch production process to flexible automation was formulated by Monahan and Smunt (1989). In that model, the sum of expected discounted costs incurred over the finite planning periods is considered as the optimisation criterion. Later, a two-stage stochastic program based on the product flexible manufacturing capacity was formulated by Fine and Freund (1990) to investigate the decision of investing in a flexible capacity compared to a dedicated capacity. These models are only relevant to capacity consideration in justifying the FMS. None of the optimisation models includes all relevant factors in the optimisation procedure. However, the research carried out by Tie and Lin (1991, 1992) had formulated an LP model which includes capital, direct, indirect and intangible costs in CIM environment. In contrast, out of all these optimisation techniques GP can be regarded as one of the most effective and suitable optimisation approache for consideration of multiple criteria in resource allocation.

6.5.1 Goal Programming

Goal programming is a powerful optimisation tool which relies on the strength of the highly developed and tested technique of LP. It provides a simultaneous solution to a complex problem of competing and diverse objectives. In addition, GP is capable of providing solutions for decision problems with single goal and multiple sub-goals, as well as problems of multiple goals and sub-goals (Lee, 1979). The fundamental approach in GP relies in establishing a specific numeric value as a target goal for each of the attributes, then formulating an objective function for each attribute, and then obtaining a solution that minimises the deviation of the weighted sum of these objective functions.

The methodology relies on minimising the unwanted deviations from those target goals. Lexicographic (or pre-emptive) GP and weighted GP are the two main approaches used in solving GP problems. In pre-emptive GP, hierarchies of goals are assigned to the objective functions of the decision problem. Then the solution algorithm sequentially solves the objective functions starting from the highest priority goals. Having satisfied the highest priority goals, the lower priority goals are attempted to be satisfied with minimal deviation from the target goals. This approach is beneficial when extensive differences exist between the hierarchy of the goals. On the contrary, in the non-pre-emptive (or weighted) GP, all the goals are given comparable priority to reach a solution. Nevertheless, the unwanted deviations can be given weights according to their relative importance and their summation be minimised.

Siha (1993) proposed a decision model based on a lexicographic GP algorithm, which incorporated tangible and intangible factors. However, the main disadvantage of this model is it can only be used to select the most appropriate system available under a particular technology and it could not be used to select a variety of appropriate technologies under the framework of CIM.

6.6 ACHIEVING OPTIMUM RESOURCE ALLOCATION

The proposed methodology for the optimal resource allocation in achieving the strategic goals of the investment provides a better and comprehensive approach compared to the existing approaches offered by various DSS to analyse investment decisions in AMT and CIM. In addition, by changing the parameters or values in the model, the model can be manipulated to reveal various scenarios due to the change in important variables of the decision problem. The GP algorithm described in this section is utilised in our proposed approach. However, optimisation procedure can be reached by two paths as shown in Figure 6.1. This procedure only requires completion of Stage 1 of the methodology to analyse the investment portfolios, since, the priority weights computed in Stage 1 are utilised for formulating the goals of the GP algorithm.

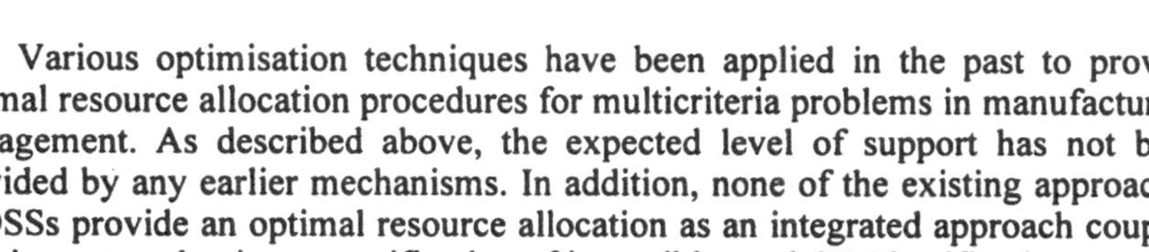

Figure 6.1 Paths to reach optimisation stage

Various optimisation techniques have been applied in the past to provide optimal resource allocation procedures for multicriteria problems in manufacturing management. As described above, the expected level of support has not been provided by any earlier mechanisms. In addition, none of the existing approaches or DSSs provide an optimal resource allocation as an integrated approach coupled with impact evaluation, quantification of intangibles and CIM justification.

In a manufacturing management decision-making exercise, all the attributes which are strategically important for the investment analysis should be given equal priority among them. Application of a linear-weighted GP model offers a facility to allocate the relative importance of each attribute on the same hierarchy level to reach an optimal solution. Hence, in this book, a linear-weighted GP model is proposed for allocating the available capital to satisfy the strategic considerations and synergy benefits. By utilising the proposed optimal resource allocation procedure, a decision-maker will gain confidence in the decision-making process quickly and effectively when solving multicriteria problems related to investments in AMT and CIM.

For the allocation of investment capital among various components of CIM, it is suggested to utilise the database consisting of percentages of improvement ratios, which could be gathered from the literature and the knowledge gained by companies that have implemented components of CIM or AMTs. A typical database, which consists of gathered improvement ratios from the available literature on AMT and CIM for all the attributes listed in this book against each component of CIM, is utilised in this approach.

Let

B_{Tij} = Percentage benefits by technology T_i for the j th attribute (from database) (where, i = 1 to n and j = 1 to m)

BL = Budget limitation of a company on new investments

I_{Ti} = Investment in technology Ti

PWj = Priority weight for attribute j

S_j^+ = Amount by which the j th goal is numerically exceeded

S_j^- = Amount by which the j th goal is numerically under achieved

The limiting constraints in the approach are formed as follows:

All the investments in components of CIM should be less than or equal to the budget constraint of a company. Hence, the limiting equation is written as:

$$\sum_{i}^{n} I_{Ti} \leq BL \tag{6.1}$$

The capital investments are positive numbers, hence

$$I_{Ti} \geq 0 \tag{6.2}$$

In general, the benefits experienced by investing in a particular technology lies between a feasible range and it can be expressed by a diagram shown in Figure 6.2. An AMT will not be able to provide large improvements in benefits related to the investment capital, beyond the maximum limit of the investment amount for the AMT. On the other hand, to utilise the benefits of an AMT, the investment should be greater than or equal to the minimum limit.

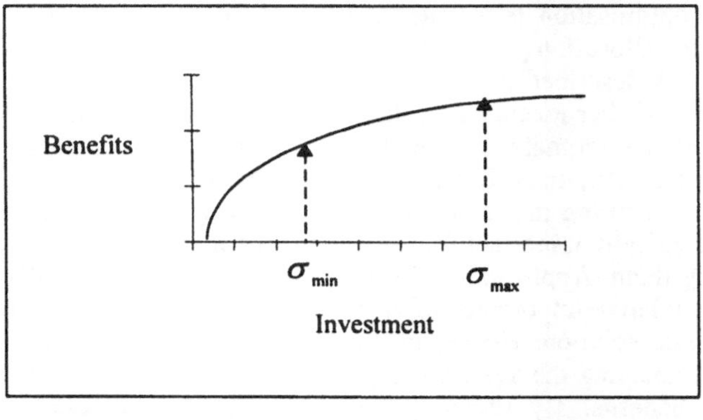

Figure 6.2 Effect of investment on achieving the benefits

The feasible investment region for a technology T_i lies between $\sigma_{min,i}$ and $\sigma_{max,j}$. Hence, the limiting equation can be formed as:

$$\sigma_{min,i} \leq I_{Ti} \leq \sigma_{max,i} \tag{6.3}$$

However, the given methodology in this book offers the flexibility of changing these limiting parameters. The limiting parameters can be entered by the decision-maker depending on the knowledge and information available.

The combination of all improvements gained by integrating various components of CIM should reach the priority weight of the strategic objective in relation to the limitation on the budget. Hence, the limiting equation for the j th goal can be written as:

$$\sum_{i=1}^{n} B_{Tij} * I_{Ti} \leq PW_j * BL \tag{6.4}$$

By inserting the deviational variables S_j^+ and S_j^- in Equation 6.4, it can be rewritten as:

$$\sum_{i=1}^{n} B_{Tij} * I_{Ti} + S_j^- - S_j^+ = PW_j * BL \tag{6.5}$$

The synergy should maximise the benefits gained by having the integration among technologies and functional units. Hence, an optimisation equation can be written as:

$$\text{Maximise } Z1 = \text{synergy by integration} = \sum_{i=1}^{n} \theta_i * I_{Ti} \tag{6.6}$$

Where, as in Equation 2.2 (Chapter two, page 30), θ_i denotes the summation of functional inter-relationship ratios for the i th AMT. This value is obtained from the functional inter-relationship matrix proposed in Chapter two (see Table 2.1, page 27). Then, Equation 6.6 can be incorporated into the goal programming by adding the deviational variables S_0^- and S_0^+ to be consistent with Equation 6.5. Then it can be formulated as:

$$\sum_{i=1}^{n} \theta_i * I_{Ti} + S_0^- - S_0^+ = PW_0 * BL \tag{6.7}$$

Where, $PW_0 = 1$

This algorithm reflects the premise that maximising the over achievement of the goals within the budget constraint will be beneficial to the company. Therefore, the GP algorithm for the resource allocation is rewritten as:

$$\text{Maximise } Z = \sum_{j=0}^{m} PW_j * S_j^+ \tag{6.8}$$

Subject to satisfying the constraints imposed by Equations 6.1, 6.2 and 6.3.

6.7 CONCLUDING REMARKS

It has been found that goal programming (GP) offers an appropriate mechanism to analyse multiple goal optimisation problems. The proposed optimisation methodology, which is based on GP, is utilised in the integrated approach to CIM justification and optimisation. The proposed methodology achieves the optimal resource allocation for the investment in AMTs, based on the strategic goals of the company and provides a better and comprehensive approach compared to the existing approaches offered by various decision support systems to analyse investment decision in AMTs as components of CIM. However, to achieve the competitive capabilities of the companies we must concentrate on long-term results rather than short-term results. By focusing on long-term results we will be able to achieve the set goals with confidence and improved capabilities. The next chapter illustrates the application of the integrated approach for CIM justification and optimisation.

6.8 REFERENCES

1. Berliner, C. and Brimson, J.A., 1988, *Cost Management for Today's Advanced Manufacturing*, Harvard Business School Press.
2. Burstein, M.C., 1986, "Finding the economical mix of rigid and flexible automation for manufacturing systems", *Proceedings of the 2nd ORSA/TIMS Conference on Flexible Manufacturing Systems: Operations Research Models and Applications*, Amsterdam, pp. 43–54.
3. Chandra, J. and Schall, S.O., 1988, "Economic justification of flexible manufacturing systems using the leontief input-output model", *The Engineering Economist*, Vol 34 No 1, pp. 27–50.
4. Fine, C.H. and Freund, R.M., 1990, "Optimal investment in product-flexible manufacturing capacity", *Management Science*, Vol 36, No 4, pp. 449–465.
5. Hess, G.J., 1992, "Beyond CIM- optimised enterprise integration", *Proceedings of Autofact '92 Conference*, USA, pp. 2–9 to 2–24.
6. Kuei, C.H., Lin, C., Aheto, J. and Madu, C.N., 1994, "A strategic decision model for the selection of advanced technology", *International Journal of Production Research*, Vol 32 No 9, pp. 2117–2130.
7. Kulatilaka, N., 1988, "Valuing the flexibility of flexible manufacturing systems", *IEEE Transactions on Engineering Management*, Vol 35 No 4, pp. 250–257.
8. Lee, S.M., 1979, "Goal programming methods for multiple objective integer programs", *American Institute of Industrial Engineers Inc.*, USA, pp. 1–27.
9. Monahan, G.E. and Smunt, T.L., 1989, "Optimal acquisition of automated flexible manufacturing processes", *Operations Research*, Vol 37 No 2, pp. 288–300.
10. Siha, S., 1993, "A decision model for selecting mutually exclusive alternative technologies", *Computers In Industrial Engineering*, Vol 24 No 3, pp. 459-475.
11. Tie, H.Y., 1992, *CIM Optimisation*, Ph.D. Thesis, University of South Australia.

12. Tie, H.Y. and Lin, G.C.I., 1991, "Towards CIM optimisation", *International Journal of Computer integrated Manufacturing Systems*, Vol 4 No 1, pp. 184–197.
13. Tie, H.Y. and Lin, G.C.I., 1992, "Incorporating CIM optimisation in manufacturing strategy", *International Journal of Computer integrated Manufacturing Systems*, Vol 5 No 4, pp. 311–325.

CHAPTER SEVEN

A Decision Support System for CIM Investment

7.1 OVERVIEW

The absence of adequate analysis and justification methods is a major problem in dealing with investment decisions for CIM and AMT. To overcome this issue a number of decision supporting tools addressing multicriteria problems have been developed during the last few decades. However, as discussed previously, an integrated approach for CIM justification has not been proposed by earlier researchers. An integrated approach should provide facilities to identify the appropriate subsystems based on the strategic considerations, evaluate the impact of subsystems company-wide and on their prime functional units, quantify intangibles and include them in the evaluation procedure, evaluate benefits due to synergy, incorporate synergy in justification, and optimise resource allocation.

As explained in Chapter three, incorporating various issues involved in justifying and optimising investments in integrated technologies puzzle manufacturing managers, since they have been justifying investments purely based on financial returns and as standalone equipment. The authors of this book realised that the complex issues in justification of CIM requires a decision-making mechanism in the form of a Decision Support System (DSS), which will be innovative and be able to assist the manufacturing managers to resolve the challenges faced in justifying CIM implementation. Therefore, the concepts explained in previous chapters have been incorporated into a DSS and developed as an innovative prototype system. This chapter illustrates, by use of an example, the application of the DSS, and the necessity to incorporate intangibles and synergy evaluation in the integration of AMTs.

7.2 NEED FOR DECISION SUPPORT SYSTEMS

CIM is a complex technology and requires huge investment. This requires a thorough understanding of fundamentals to make correct investment decisions. The rewards and benefits of CIM are extensive, but wrong decisions will lead to disastrous results with minor or no chance of recovery. This phenomenon arises due to large-scale initial investment and far reaching organisational changes required for the migration to CIM from conventional technologies. Hence, when formulating CIM investment policies, management must be mindful of the importance and dimensions of the decisions and strategies. A capital-intensive project such as CIM, which has many parameters to consider, usually requires a detailed analysis with many intricate procedures. All these intricate analyses can be provided by a DSS with simplicity and in a manner easy to understand. In addition, to accommodate the conflicting goals and provide a balanced approach, as a holistic objective to the company, a DSS is indispensable to a decision-maker.

Today, we have computer hardware and software that are very advanced and powerful in processing, mathematical modelling and analysis. Even though, computers are merely rapid and accurate mathematical and information processing units and do not understand the decision-making criteria or the concepts of strategic benefits, the application of computers as an analytical tool will result in more accurate result in less time. Manufacturing planners deal with many alternatives in many areas of management decision-making. They include sales order processing; forecasting; aggregate production scheduling; distribution of finished goods, raw materials, work-in-process inventory; master production scheduling; material requirement planning; capacity requirement planning; bills of materials; inventory; and quality control. Since all these decision-making areas require mathematical models, the use of Decision Support Systems (DSSs) in a CIM environment is a necessity for effective information processing and to resolve complex analysis.

In today's business world, most of the managers and engineers are provided with powerful computers for their use. Hence, the development of a microcomputer based AMT evaluation tool, could bring the evaluation of an AMT to the fingertips of the manager as an efficient and affordable tool. The use of DSS as a computer oriented application will allow the results to be well recorded and give more time for a decision-maker to analyse issues which cannot be handled by a computer. These features and opportunities provided by computer applications have motivated many researchers to develop DSSs for the justification of CIM. In addition, the use of DSSs in solving multicriteria problem helps the decision-makers in reaching a solution rapidly, since the computer performs all the complex calculations and logical formulations, leaving the decision-maker only to construct the problem in a correct manner.

7.3 DECISION SUPPORT SYSTEMS

A decision support system is generally defined as an interactive Computer-Based Information System (CBIS) that helps decision-makers to solve semi-structured and unstructured problems through an extensive use of a database and mathematical and logical models. A DSS allows a decision-maker to combine human judgement with a meaningful computer output to derive the right decision for the problem. The DSS is a subclass of CBIS. The concept of DSS was originally stated by Gerrity (1971), as being "An effective blend of human intelligence, information technology, and software that interact closely to solve complex problems". A DSS helps a strategic planner in evaluating the effect of AMT on the performance of the organisation and determining the parameters which affects the costs and benefits of such a system. By utilising a DSS, an individual or group can benefit by making constructive and informed decisions in reduced time.

The output of a CBIS is utilised in assisting management in planning and controlling activities. The information required for management functions may fall into strategic, tactical and operational levels of activity depending on the nature of the problem and the requirement of the management. Strategic information is useful for senior management where planning and decision-making is performed. Tactical information and operational planning information for the near future is dealt with by middle management. The lower management, responsible for daily transactions and operations of the plant, is concerned with operational information. Depending on the level of information required the classification of the CBIS would vary. Some classifications of computer based information systems (CBISs) are explained below. However, the description of these CBISs may overlap one another to varying degrees, since there is no standard definition for these CBISs and the definition may vary depending on the application.

7.3.1 Transaction Processing Systems

Transaction processing systems (TPS) are mainly used for data collection, storage, and processing of summaries of reports for operational-level personnel and for input to other types of information systems (e.g. shop floor data collection systems, payroll systems). TPS represent application of a CBIS for routine and ordinary business transactions.

7.3.2 Management Information Systems

Management information systems (MIS) are designed to assist the management functions and to improve the performance of the operations of the enterprise. MIS provide information related to planning, organising and controlling the resources of an enterprise. The decision support provided by MIS has a distinct focus and it is provided to middle management, usually allowing ad hoc inquiry and report generation using a database for structured problems which are often programmable (e.g. financial or accounting reports).

7.3.3 Expert Systems

Expert systems (ES) are highly specialised, decision modelling tools that are designed to solve certain problems in particular areas. In these types of systems, expertise or knowledge is transferred from an expert or a few experts via a computer to a non-expert, by storing and retrieval. This expertise is coded as a series of rules and allows that knowledge to be applied to specially structured problems by a less experienced person as it would have been handled by an expert. Capturing the expertise of the experts allows the experts to expand their activities beyond the routine tasks and permits the knowledge to be stored for future use. There are many ES available for applications in the manufacturing environment. These include process planning, assembly operations, facility layout, scheduling, and material handling.

7.3.4 Decision Support Systems

A decision support system, supports decision-making but it does not replace the managers in making a decision as an expert system. A DSS often includes modelling and analysis capabilities, forecasting and statistical techniques, data management, communication links, reporting, graphics and a user-friendly environment. A DSS provides technology to cope with the diversity and complexity of the information involved and helps decision-makers solve semi-structured and structured problems with managerial judgement.

As explained in Chapter four, the authors of this book have developed a DSS for CIM investment (DSSCIM). The application of the DSSCIM is explained as an example in Section 7.5. Section 7.4 gives the reader with the background information on the DSSCIM and the assumptions used in the example.

7.4 BACKGROUND AND ASSUMPTIONS

The implementation of CIM requires correct forward thinking and appreciation of AMTs. The application of the AMTs in manufacturing industries has resulted in significant improvement towards productivity, quality and competitiveness. However, some strategic questions are often raised by management regarding the implementation of new technologies. They are:

1. Which business functions will benefit from new technologies?
2. Can different technologies be integrated into a true CIM environment?
3. If the answer to Question two is yes, what are the combinations of AMTs to optimise the investment?
4. Can an adequate return of investment be ensured in capital intensive investments?

The DSSCIM, developed by the authors, assists manufacturing managers to determine the answers to the above questions. The first question is addressed in Stage 3 of the system, however, it is carried through from Stage 1 to Stage 3 (see Chapter four). The second question is outlined in Stage 4 of the DSSCIM. The third question is addressed in Stage 6 of the system. Stage 5 of the research project presents a technique, which provides a facility to evaluate the return of investments. Goal optimisation is addressed in Stage 5 and it ensures an optimal balance of technologies is selected. In addition, the DSSCIM enhances the understanding of synergy among technologies and functional units by incorporating a Functional Inter-Relationship Matrix (FIRM) in the methodology and utilising synergy functions for evaluating investments.

The rationale for AMT justification must be consistent, so that decision-makers can evaluate their investments under a similar environment. In addition, any real investment analysis involves considerations of reasonable and realistic assumptions. In the illustrative example, the following assumptions are made:

1. Input data for the comparison process is available from senior management. The additional data for cost evaluation is available from various departments in the company;
2. The initial inventories of raw materials and finished goods are at zero level;
3. The capital for investment is entirely from the resources held by the company as a single source and is not borrowed from any other sources;
4. The capital depreciation associated with the system includes all the peripherals associated with that system;
5. The salvage value of a system and its book-value at the end of the planning period is equal;
6. The planning horizon is less than or equal to 10 years; and
7. Inflation is zero during the entire calculating period.

The software environment of the prototype DSSCIM was developed on Microsoft Excel version 5.0c running under windows for work-groups version 3.11. The Visual Basic for Application, which is embedded with the Microsoft Excel, is utilised in controlling the necessary sequences and the analysis of the DSSCIM.

7.5 APPLICATION OF DSSCIM

In order to illustrate the methodology and the application of the DSSCIM by an example, a hypothetical company which is interested in investing in technologies that are suitable to achieve the strategic concepts of the company, is considered.

The example sequentially takes the decision-maker through all the stages of the analysis procedure in the DSSCIM. Let a manufacturing company which has an annual total operating cost of 25 million dollars, decide to invest about $3 million in AMTs in order to compete with others in the market. The decision-makers of the company have decided to use the DSSCIM to justify the investments and select the AMTs based on certain strategic considerations. When the application is activated, the user is presented with an introductory screen as shown in Figure 7.1.

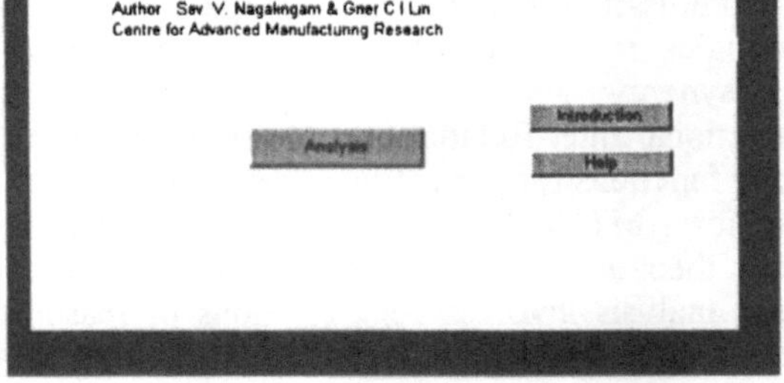

Figure 7.1 Introductory screen of the DSSCIM

7.5.1 Strategic Decision Profile

The purpose of the buttons depicted on the introductory screen is as follows:

- Button named " Analysis" activates the initiation process of DSSCIM;
- Button named " Help" activates a work sheet called Help.xls, which contains information on help, in the same directory; and
- Button named " Introduction" activates a work sheet called Dssslide.xls that contains in-build slides for presentation.

Selection of the " Analysis" button presents the user with the attributes of the strategic decision profile as shown in Figure 7.2.

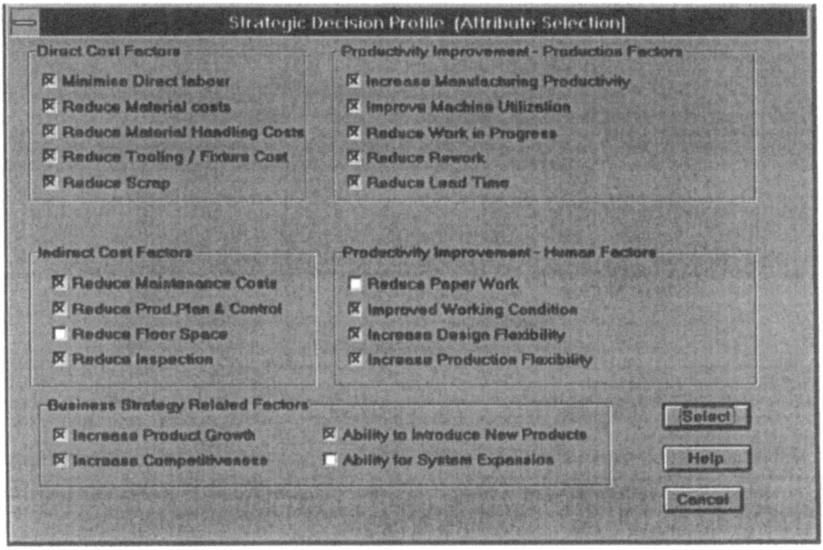

Figure 7.2 A dialog box depicting the attributes of the strategic decision profile

The decision-makers determined that the following attributes among those given in each category are important for the company for the consideration of the investment analysis:

- Category 1: Direct Cost Reduction (DCR)
  This category includes: attributes such as Minimise direct labour; Reduce material costs; Reduce material handling costs; Reduce tooling/fixture costs; and Reduce scrap.
- Category 2: Indirect Cost Reduction (ICR)
  This category includes: attributes such as Reduce maintenance costs; Reduce production planning and control costs; and Reduce inspection costs.
- Category 3: Productivity Improvement Factors – Production Related (PIPF)

This category includes: attributes such as Increase manufacturing productivity; Improve machine utilisation; Reduce work in progress; Reduce rework; and Reduce lead-time.

- Category 4: Productivity Improvement Factors – Human Related (PIHF)
  This category includes: attributes such as Improved working conditions; Increase design flexibility; and Increase production flexibility.
- Category 5: Business Strategy (BS)
  This category includes: attributes such as Increase product growth; Increase competitiveness; and Ability to introduce new products.

Once the desirable strategic attributes are identified and selected for the analysis on the dialog box shown in Figure 7.2, the DSSCIM prompts the user, as shown in Figure 7.3, to select the type of analysis in order to rank the selected attributes.

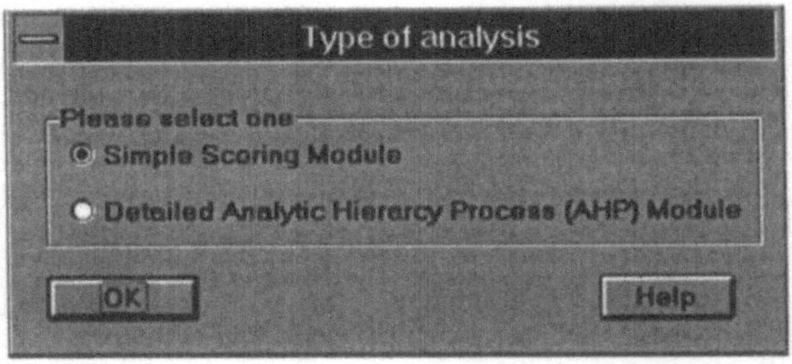

Figure 7.3 A dialog box requesting the selection for the type of analysis

The decision-makers decide to evaluate the attributes by the AHP method, since it gives a structured approach to prioritise the preferences of the company and assign weights accordingly. Then the DSSCIM prompts the user to prioritise the first level of the hierarchy (that is prioritising the categories shown in Figure 4.2 of page 60, by the dialog box shown in Figure 7.4 (page 105). Preferences of the paired comparisons assigned by the decision-makers are shown in Table 7.1.

Table 7.1 Matrix of paired comparisons at the first level of the hierarchy

Criteria Comparison	DCR	ICR	PIPF	PIHF	BS
Direct Cost Reduction [DCR]	1.00	3.00	3.00	5.00	0.33
Indirect Cost Reduction [ICR]	0.33	1.00	0.33	1.00	0.20
Productivity Improvement (Production related) [PIPF]	0.33	3.00	1.00	3.00	0.33
Productivity Improvement (Human related) [PIHF]	0.20	1.00	0.33	1.00	0.20
Business Strategy [BS]	3.00	5.00	3.00	5.00	1.00

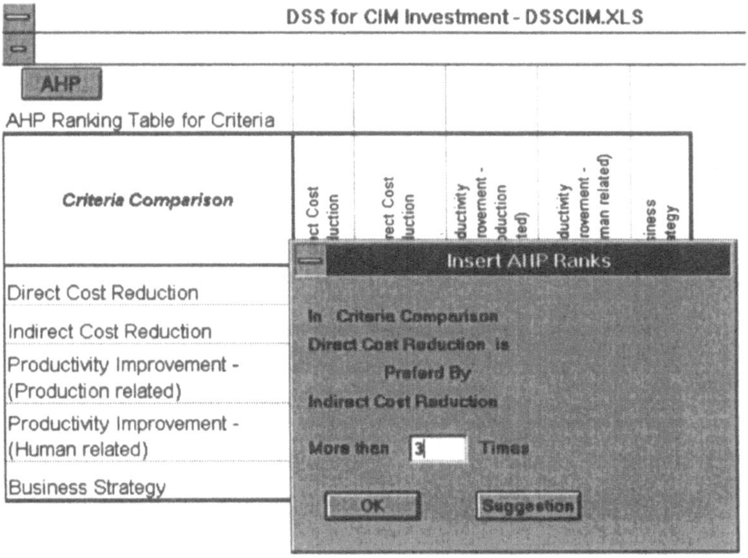

Figure 7.4 A dialog box requesting the prioritisation of the categories

Once these pair-wise comparisons are assigned, the DSSCIM calculates the consistency ratio for the indices assigned and display it to the decision-maker. When the decision-makers have accepted that value, the DSSCIM displays a dialog box, as shown in Figure 7.5, to select the second level of the categories and continue the prioritisation among them.

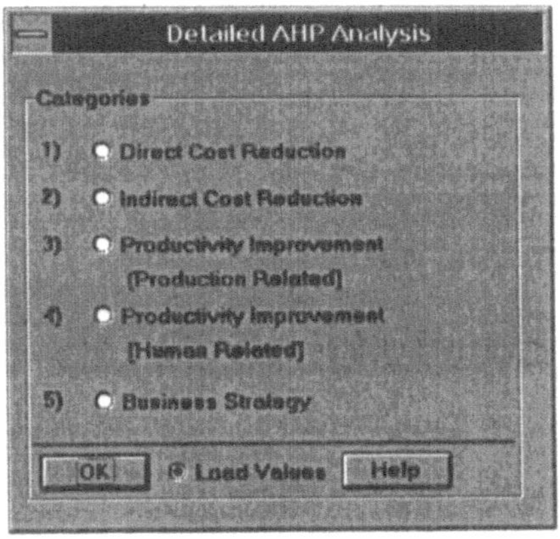

Figure 7.5 A dialog box requesting the user to select categories

For all the categories on the second level of the hierarchy, when the selection is made on the dialog box of Figure 7.5, the dialog box shown in Figure 7.4 (page 105) appears and the associated worksheet is activated. Then the consistency ratio is computed by the DSSCIM and is displayed to the users for verification. The indices assigned by the decision-makers for the first category is shown in Tables 7.2. For other categories, similar tables can be constructed and the values can be assigned.

Table 7.2 Matrix of paired comparisons of DCR at the second level of hierarchy

Direct Cost Reduction	Direct labour	Material Costs	Material handling costs	Tooling and Fixture Costs	Scrap Costs
Direct labour	1.00	0.33	0.20	0.20	0.11
Material Costs	3.00	1.00	0.33	0.20	0.14
Material handling Costs	5.00	3.00	1.00	0.33	0.20
Tooling/Fixture Costs	5.00	5.00	3.00	1.00	0.20
Scrap costs	9.00	7.00	5.00	5.00	1.00

Consistency Ratio = 0.087

7.5.2 Evaluation of Performance Measures

On completion of assignation of pair-comparisons at the second level of the hierarchy, the user is requested to select the "Load Values" option shown in Figure 7.5 (page 105) in order to compute the third level of the hierarchy. The dialog box requesting the user to assign indices for the third level is shown in Figure 7.6. The assigned indices for the third level are shown in Table 7.3 (page 107). As explained in Chapter four, this third level is considered as the performance measure evaluation stage for investing in AMTs and in traditional systems. Performance measures are determined on the basis of the rating provided by the decision-makers, for the automation alternative with respect to each criterion, either with the status quo or with traditional systems.

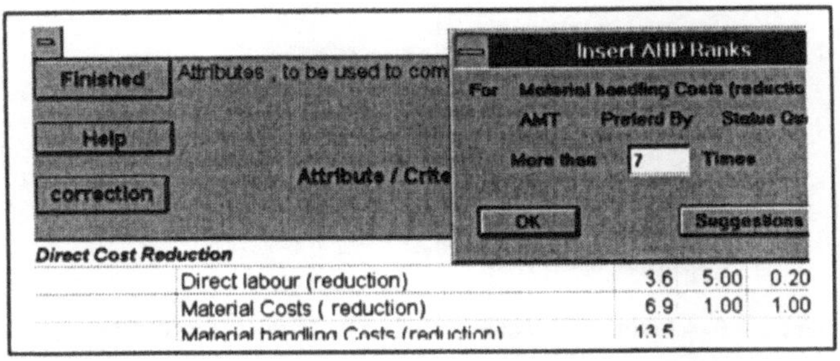

Figure 7.6 A dialog box requesting the user to assign the third level of the hierarchy

Table 7.3 Performance measures at the third level of the hierarchy ranking

Attribute/ Criterion	Weight (from AHP)	Ranking AMT	Ranking Status Quo
Direct Cost Reduction			
Direct labour (reduction)	3.6	5.00	0.20
Material Costs (reduction)	6.9	1.00	1.00
Material handling Costs (reduction)	13.5	7.00	0.14
Tooling/Fixture Costs (reduction)	22.2	5.00	0.20
Scrap (reduction)	53.8	3.00	0.33
Indirect Cost Reduction			
Maintenance Costs (reduction)	10.6	0.20	5.00
Prod. plan. & control (reduction)	63.3	3.00	0.33
Inspection (reduction)	26.1	7.00	0.14
Productivity Improvement (Production Factors)			
Manufacturing productivity (increase)	35.3	5.00	0.20
Machine utilisation (improve)	6.4	5.00	0.20
Work in progress (reduce)	14.8	7.00	0.14
Rework (reduce)	10.1	3.00	0.33
Lead time(reduce)	33.5	5.00	0.20
Productivity Improvement (Human Related Factors)			
Working conditions (improve)	14.3	5.00	0.20
Design flexibility (increase)	42.9	7.00	0.14
Production flexibility (increase)	42.9	9.00	0.11
Business Strategy			
Increase Product Growth	63.3	3.00	0.33
Increase competitiveness	26.1	7.00	0.14
Ability to introduce new products	10.6	9.00	0.11

Three buttons, "Correction", "Help", and "Finished" build on the worksheet (see Figure 7.6, page 106). On completion of performance measure comparison, the user is requested to activate the button "Finished" as in Figure 7.6. The purpose of the "Correction" button is to make corrections to any unsatisfactory computation of performance measures. If the performance measurement is satisfactorily achieved, the DSSCIM indicates by displaying a message dialog box. Subsequently, an alert message is given by the DSSCIM, before proceeding to the third stage of the analysis, to indicate it is generating suggested subsystems for investment. The suggested subsystems are generated according to the database and a rule-based algorithm and it is displayed, as it is shown in Figure 7.7 (page 108). By contrast, when the performance measure is unsatisfactory, the DSSCIM requests the user to reassign the weights at the third level.

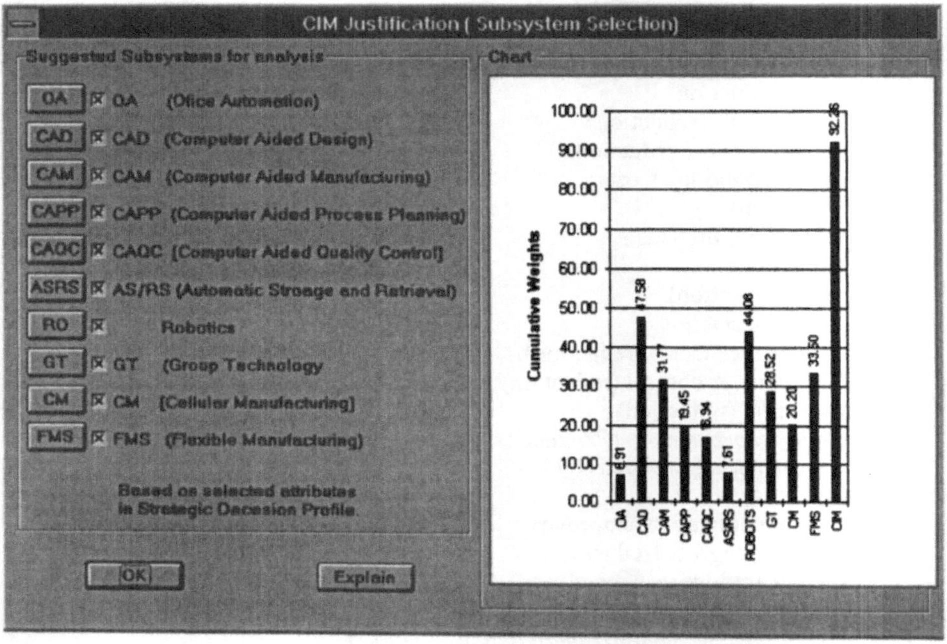

Figure 7.7 A dialog box showing the suggested subsystems

In this case, the decision-makers decide that the AS/RS and CM did not have sufficient priority and need not be installed with the investment of FMS. Once the selection is made, the DSSCIM displays the FIRM, which consists of the technologies selected from the dialog sheet (see Figure 7.7), and also the functional units of the company in order for the decision-makers to assign the ratios of the functional inter-relationships. This information is relayed to the decision-maker by a message dialog box.

7.5.3 Identification of Functional Inter-Relationships

The methodology for assigning values on the FIRM is explained in Chapter two. As explained, the ratios can be assigned either by the structured approach provided in Chapter two or by subjective estimation based on experience. The ratios assigned by the decision-makers are shown in Table 7.4 (page 109).

Abbreviations used in Table 7.4 are as follows:

Admin:	Administration	QAC:	Quality Assurance and Control
D & E:	Design and Engineering	W & S:	Warehousing and Shipping
Manu:	Manufacturing	Asy:	Assembly

PPC: Production Planning and Control R & D: Research and Development

Table 7.4 Assigned ratios of FIRM

Technologies	Functional Units							
	Admin	D & E	Manu	PPC	QAC	W & S	Asy	R & D
Office Automation	1.00	0.01	0.05	0.01	0.01	0.20	0.00	0.20
Computer Aided Design	0.05	1.00	0.30	0.20	0.02	0.06	0.03	0.40
Computer Aided Manufacturing	0.01	0.10	1.00	0.40	0.40	0.01	0.20	0.20
Computer Aided Process Planning	0.01	0.01	0.03	1.00	0.20	0.08	0.20	0.20
Computer Aided Quality Control	0.01	0.01	0.05	0.07	1.00	0.01	0.01	0.01
Robotics	0.00	0.04	0.60	0.05	0.10	0.02	1.00	0.01
Group Technology	0.00	0.01	1.00	0.20	0.10	0.20	0.30	0.20
Flexible Manufacturing System	0.07	0.20	1.00	0.60	0.50	0.20	0.30	0.20

7.5.4 Evaluation of Tentative Cost-Impact

Having assigned the ratios on FIRM, the decision-maker is requested to activate the button "OK" built on the first row of the FIRM. Once it is completed, the user is asked by displaying a dialog box to verify whether they wish to perform the tentative cost-impact analysis. When the answer is affirmative, then the DSSCIM displays the cost-share dialog box, as shown in Figure 7.8 (page 110), to determine the total operating cost of the company and the share of cost in each functional unit.

When the total share of costs is up to 100 per cent, the user is by a dialog box asked to select the type of analysis for the tentative cost-impact. As shown in the flowchart of the DSSCIM (see Figure 4.1, page 55), the DSSCIM gives two options such as detailed analysis of attributes and analysis by system. However, if the total share is over 100 per cent, the user is informed by a message box that the total share of costs is unsatisfactory and asked to ensure that the assigned ratios are satisfactory.

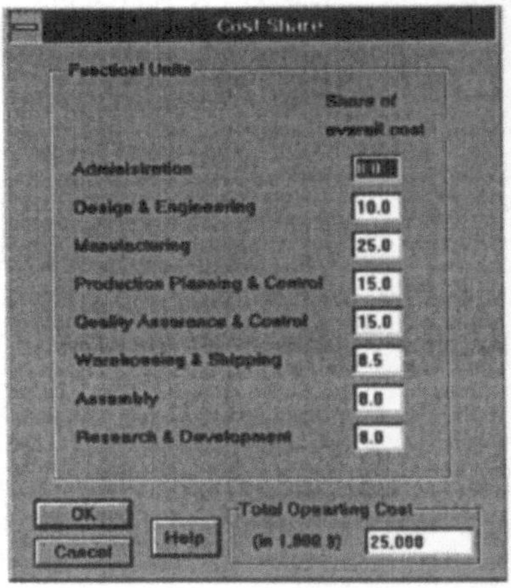

Figure 7.8 Dialog box in determining the share of cost

In this exercise, the decision-makers of this hypothetical company (we) decided to perform the tentative analysis, using the "Detailed analysis of attributes" option, since it will give us a sufficient learning experience in realising the impact of the attributes on each functional unit. Once the selection is made, the DSSCIM displays an informative dialog box asking us to select the subsystems in the subsequent dialog box to analyse the attributes in detail. Then it displays a dialog box (Figure 7.9) to select the analysis for each AMT. However, if the user has chosen the system-wide analysis option, a different dialog focusing only on the holistic view of the department, where the AMT was assigned, would be displayed by the DSSCIM and would have guided us to assign values for all the selected departments.

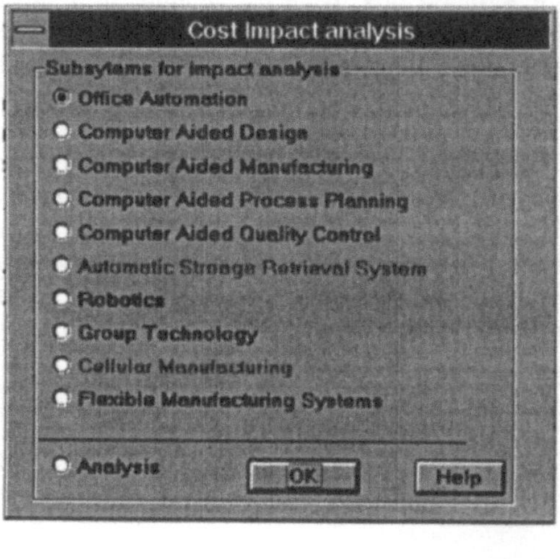

Figure 7.9 A dialog box guiding the user in performing the impact analysis

As explained in Chapter five (page 71), once the selection is made, the DSSCIM selects only those attributes associated with the selected technology from among those selected in the strategic decision profile. The associated attributes and technologies are given in Appendix 2. The impact evaluation performed by the decision-makers for the office automation is shown in Table 7.5. Similar impact evaluations are performed for all AMTs under consideration. The DSSCIM only allows the selection of technologies which were pre-selected by the user from the dialog box shown in Figure 7.7 (page 108).

Table 7.5 Impact evaluation on administration department

Prime Functional Unit (Office Automation)	Administration (Effect %)
Minimising direct labour	2
Reducing lead-time	2
Reducing paper work costs	5
Improving working conditions	5
Impact within the Functional Unit	14

Once the impact evaluation is completed for all the pre-selected technologies, the DSSCIM computes the overall effect company-wide and all the functional units, and displays the results as shown in Figures 7.10 and 7.11.

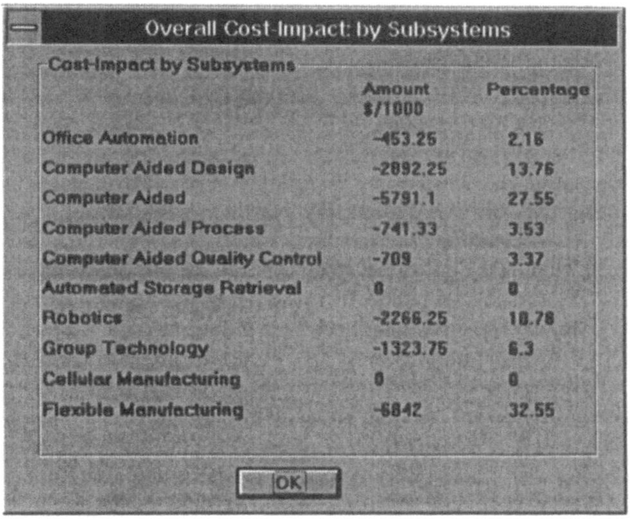
Overall Cost-Impact by Subsystems

Cost-Impact by Subsystems

	Amount $/1000	Percentage
Office Automation	-453.25	2.16
Computer Aided Design	-2892.25	13.76
Computer Aided	-5791.1	27.55
Computer Aided Process	-741.33	3.53
Computer Aided Quality Control	-709	3.37
Automated Storage Retrieval	0	0
Robotics	-2266.25	10.78
Group Technology	-1323.75	6.3
Cellular Manufacturing	0	0
Flexible Manufacturing	-6842	32.55

OK

Figure 7.10 Overall cost impact by AMTs

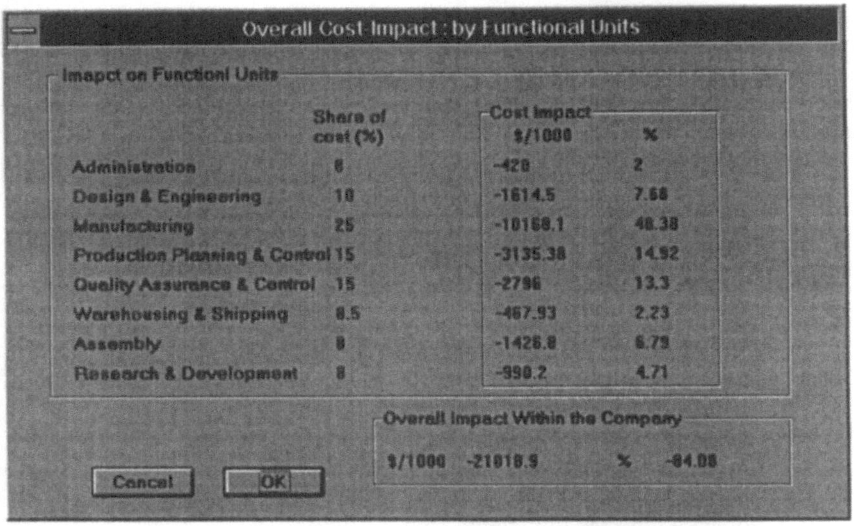

Figure 7.11 Overall cost impact by functional units

7.5.5 Evaluation of Financial Measure

Having displayed the results for the tentative cost-impact evaluation, the DSSCIM prompts the user with an informative dialog box advising them to select the appropriate subsystems from the dialog box for the financial analysis. Then it verifies whether the decision-maker is interested in the quantification of intangibles by displaying a dialog box. The decision-maker of this hypothetical company is interested in utilising the Cost-Value method in analysing the intangibles. Then the DSSCIM displays a dialog box for financial analysis, which is similar to the dialog box shown in Figure 7.9 (page 110), to guide the decision-makers in selecting the appropriate template for financial analysis.

An illustrative example in Section 7.5.5.1 is given for the Cost-Value analysis by considering the introduction of robotics in the manufacturing process. Robots are used in a wide variety of manufacturing applications. As explained in Chapter one, robots are generally regarded as machines that replace people, and they have been justified on that basis in the past. Considering the robot as a replacement for human labour limits many opportunities brought by robots while creating discontent among the workforce.

Once "Robotics" is selected from the option button in the dialog box for financial analysis, the DSSCIM opens the necessary template to analyse investments in robots and displays a dialog box (shown in Figure 7.12), to estimate the analysis period.

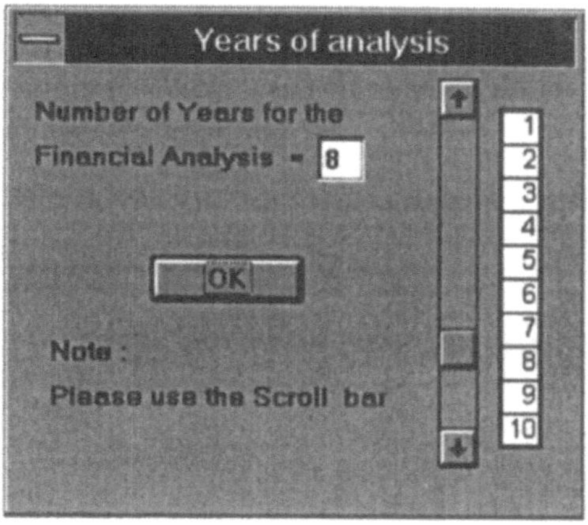

Figure 7.12 A dialog box requesting a specified period for analysis

7.5.5.1 Illustrative example for value assignment

Investment in robotics can generate many benefits to a company. As given in Appendix 2, the attributes which were associated with robots and belong within the selection made by the decision-maker on the strategic decision profile are:

- Minimising direct labour
- Reducing material handling costs
- Reducing scrap cost
- Reducing inspection cost
- Increasing manufacturing productivity
- Improving machine utilisation
- Reducing rework costs
- Reducing lead-time
- Improving working conditions
- Increasing competitiveness

Each template for financial analysis is provided with buttons such as "Display benefits", "Savings", "Value", "Calculate" and "Next AMT". When the "Savings" button on the financial sheet is activated, for each attribute DSSCIM generates appropriate dialog boxes. When the "Estimate" button is activated on that dialog box, the DSSCIM provides a series of dialog boxes based on equations derived in Chapter five (Section 5.3.1), for estimating the benefits obtainable from those attributes. Two of the dialog boxes displayed by the system are shown in Figures 7.13 and 7.14. Similar attributes are generated by the DSSCIM for all other attributes.

Attribute 1: Minimising direct labour costs

The savings from the reduction of direct labour can be estimated with a high degree of accuracy. The DSSCIM provides a dialog box as shown in Figure 7.13 for the attribute concerning minimising direct labour costs. As explained above, when the user selects the "Estimate" button, it shows a dialog box (Figure 7.14), which requests the parameters to compute the estimated benefits according to Equation 5.13 (page 75). Having assigned the parameters, the DSSCIM computes the estimated benefit and shows the calculated result as a dialog box. The decision-makers can accept that amount or substitute another amount, which satisfies them.

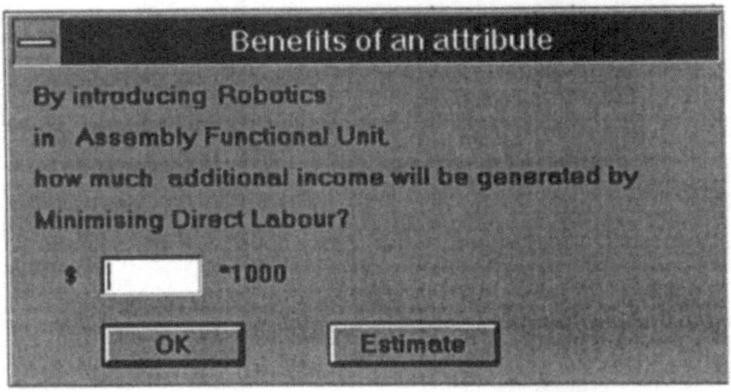

Figure 7.13 A dialog box requesting the estimation of savings that could be accrued from reductions in direct labour costs

According to Equation 5.13, the calculation is as follows, and is displayed in Figure 7.14.

Existing labour units = ELU/no = 20;
Wages/labour = $WpL = 40,000
Hence, Savings = ELU * WpL * percentage savings from the database/100
= 20 * 40,000 *70/100 = 560,000

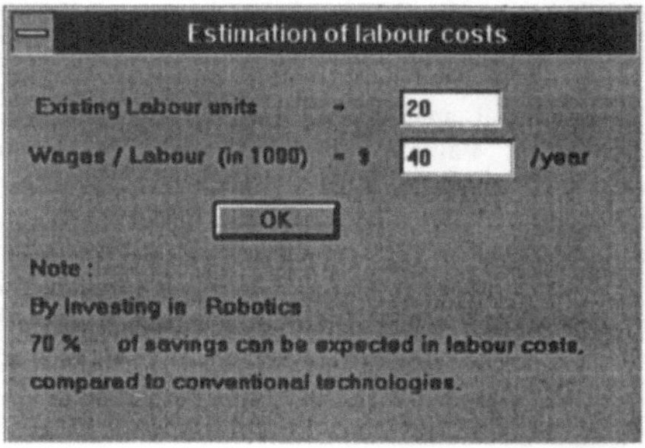

Figure 7.14 A dialog box requesting parameters to compute benefits according to Equation 5.13

Attribute 2: Reducing material handling costs
The savings from reductions in material handling costs cannot be estimated with a good degree of accuracy, since the process involves direct and indirect labour costs. However, a marginal estimation can be made on the indirect costs of material handling. For the computation, the DSSCIM displays a dialog box requesting the estimation of savings that could be accrued from reduction in material handling costs. Then a dialog box requesting parameters to compute benefits according to Equation 5.17. These dialog boxes are not shown in this chapter.

According to Equation 5.17, the calculation is as follows:
Existing material handling cost = $EMHC = 10,000
Hence, Savings = EMHC * percentage savings from the database/100
= 10,000 * 30/100 = 3,000

In order to avoid a double counting error (an error of counting same benefits or costs twice), a lower amount for indirect labour is considered, since the direct labour component has been incorporated earlier in the previous attribute estimation. Then the estimated benefit of $3,000 is displayed in a dialog box for verification.

Attribute 3: Reducing scrap
The reduction in scrap due to investing in robots can be estimated with a fair degree of accuracy. A dialog box assists the decision-makers in computing the savings according to Equation 5.20 (page 77).

According to Equation 5.20, the calculation is as follows:
Existing scrap cost = $ESC = 10,000
Hence, Savings = ESC * percentage savings from the database/100
= 10,000 * 70/100 = 7,000
Then, a dialog box with estimated savings of $7,000/annum is generated by DSSCIM and is displayed to the users.

Attribute 4: Reducing inspection costs
Reduction in the cost of inspection can be roughly estimated. Similar to those shown in Figures 7.13 and 7.14, a dialog box which computes savings according to Equation 5.28 (page 78) and a dialog box with estimated saving of $6,000/annum are shown by the system.

According to Equation 5.28, the calculation is as follows:
Existing inspection costs = $EIC = 30,000
Hence, Savings = EIC * percentage savings from the database/100
= 30,000 * 20/100 = 6,000

Attribute 5: Increasing manufacturing productivity
The capability to increase productivity will definitely place the company in a better competitive position. As above, a dialog box which computes the savings according to Equation 5.30 (page 79) is shown to the decision-maker. For this attribute also, in order to avoid the double counting error, only the production

costs of the existing machine is considered, since the labour component has been incorporated in the previous attribute estimations. Once the computation is complete a dialog box with estimated savings of $23,400/annum is generated by the system.

According to Equation 5.30, the calculation is as follows:
Existing throughput/year = ETP = 8000,
Unit production cost per unit = UPC = $4.5
Hence, Savings = ETP * UPC * percentage savings from the database/100
= 8000 * 4.5 * 65/100 = 23,400

Attribute 6: Improving machine utilisation

The capacity to improve machine utilisation will be beneficial for the company in the near future. As above, a dialog box which computes the savings according to Equation 5.31 (page 79) is shown to the decision-maker, then a dialog box with estimated amount of $120,000/annum is shown.

According to Equation 5.31, the calculation is as follows:
Existing idle time/week = EIT /hour = 10
Operating cost/hour = $OC =3
Number of hours used/week = NU = 100
Henc, Savings = EIT * OC * NU * 50 * percentage savings from the database/100
= 10 * 3 * 100 *80/100 * 50= 120,000

Attribute 7: Reducing reworking costs

This attribute can be objectively estimated within a fair degree of accuracy. A dialog box requesting parameters to compute saving in rework related costs according to Equation 5.33 (page 79) is shown by the DSSCIM.

According to Equation 5.33, the calculation is as follows:
Existing rework costs = $ERC = 30,000
Hence, Savings = ERC * percentage savings from the database/100
= 30,000 * 75/100 = 22,500

Once, this calculation is computed, a dialog box with the estimated savings of $22,500/annum is shown displayed by the DSSCIM for verification from the decision-maker.

Attribute 8: Reducing lead-time

The "Estimate" button on the NPV analysis sheet (see Table 7.6, page 118) brings up a dialog box requesting parameters to compute saving by the reductions in lead-time according to Equation 5.35 (page 80). In order to avoid the double counting error, only the production costs of existing machinery is considered, since the labour component has been incorporated in previous attribute estimations. On completion of the calculation, a dialog box with the estimated value of $240,000/annum is shown by the DSSCIM.

According to Equation 5.35, the calculation is as follows:
Existing lead time of a product = ELT/hour = 400

Operating cost/hour = \$OC = 3
Number of items produced/week = NIW = 10
Hence

Savings = ELT * OC * NIW * 50 * percentage savings from the database/100
= 400 * 3 * 10 * 50 * 40/100 = 240,000

Attribute 9: Improving working conditions by eliminating unpleasant and hazardous tasks
Hazardous tasks increase the opportunities for more accidents in the working environment and illness for the employees. Hence an improvement in working conditions will result in greater productivity and in less interruption to work. This feature can add value to the manufacturing operations and can be recovered as savings from reduced labour compensation. In addition, improvements in working conditions will result in improved morale of the employees. This feature will facilitate retention of the workers, thus avoiding higher worker turnover and retraining costs. Furthermore, it will help attract more highly qualified employees for that company. To estimate the value of the attribute, a dialog box requesting an estimated value for improvements in working conditions is provided to the decision-maker. In this example, the value of the attribute is estimated to be \$20,000.

Attribute 10: Improving competitiveness
Improvements in product quality, and having consistent quality throughout the range of products, will help the company compete successfully with its competitors. Similar to above to estimate the value of the attribute, a dialog box requesting an estimated value is displayed to the decision-maker. In this exercise, value of this attribute is estimated as \$30,000.

7.5.6 Calculation of Financial Indices

A financial evaluation template is shown in Table 7.6 (page 118). Once the value and benefit estimations are completed, the DSSCIM assigns benefits for all the evaluation periods within the study period with an estimated learning curve effect. Then the decision-makers are requested to select the "Calculate" button on the financial analysis sheet to continue the calculation for the financial indices for the investment.

	A	B	C	D	E	F	G	H	I	J	K	L
1	*Technology* Display benefits	*Robotics*										
2	*Prime Functional Unit*	*Assembly*										
3	***Year***	[redacted]										
4	***Cost- Value analysis incorporated***											
5		(in 1000s)										
6	**Investment, Capitalized**											
7	Equipment Cost	1200										
8	Accessories & Tooling Cost	30										
9	other---	8										
10	Total CapitalInv	1238										
11	Tax Credit for the investment	10.00%										
12	Investment Tax Credit	123.8										
13	Effective Cap.Invesment	1064.2										
14												
15	**Investment, Expended**											
16	Engineering	35										
17	Installation	20										
18	Startup	10										
19	Other	3										
20	Total ExpensedInv.	68										
21												
22	**Operating Savings (Costs)**											
23												
24	Saving **Robot** Value		981.9	1001.538	1021.569	1042	1062.84	1084.097	1105.779	1127.894		
25	other Savings											
26	Other Costs											
27	Total Operating Savings	-68	981.9	1001.538	1021.569	1042	1062.84	1084.097	1105.779	1127.894	0	0
28												
29	**Analysis**	Calculate										
30												

Table 7.6 Template for evaluating financial analysis

Then the DSSCIM presents the user with three options of depreciation methods, as shown in Figure 7.15. When the depreciation method is selected, the DSSCIM displays a dialog box, as shown in Figure 7.16, to determine the depreciation parameters for the selected method.

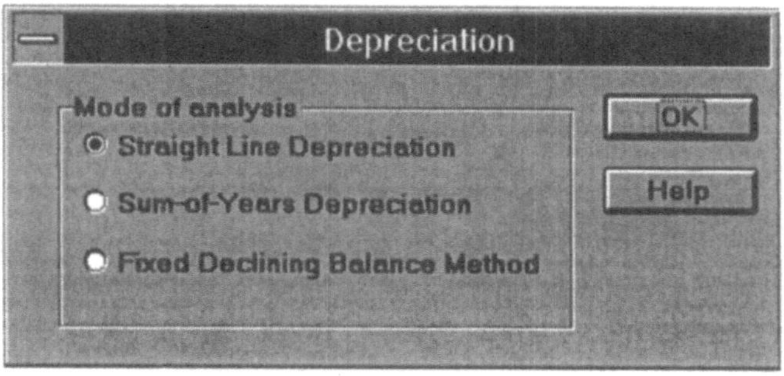

Figure 7.15 A dialog box requesting the selection of a depreciation method

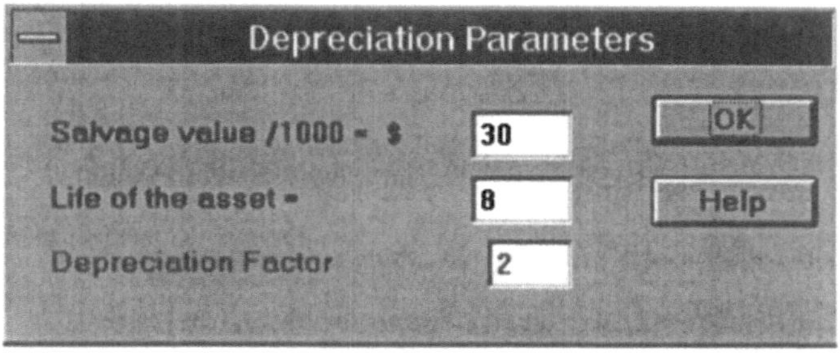

Figure 7.16 A dialog box requesting depreciation parameters

Once the parameters required to complete the depreciation calculation have been assigned, the DSSCIM computes the net-cash flow and requests the user to assign the appropriate corporate tax rate applicable to the company, as shown in Figure 7.17 (page 120). The dialog box shown in Figure 7.17 is repeated with dynamic labels displaying the cash-flow and the period under investigation for all financial periods within the study in order to ascertain the appropriate company tax rate for each period.

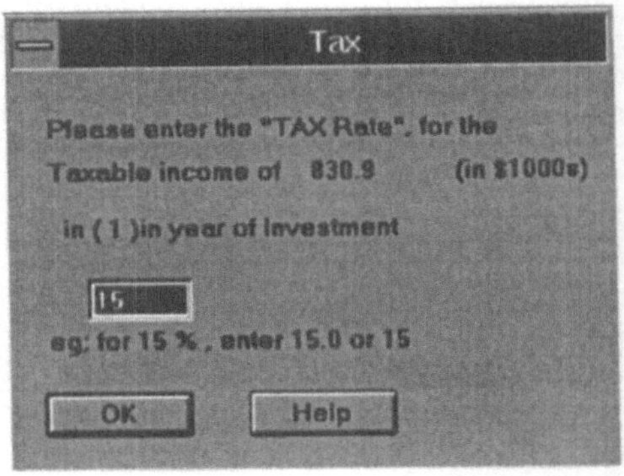

Figure 7.17 A dialog box requesting the tax rate for year 1

Once the assignation of tax rate is completed for every financial year within the study period, the DSSCIM alerts the user that the next stage of computation is to assign a MARR for calculating the NPV of the investment, by means of a dialog box. This is provided to indicate a change in the sequence, then a dialog box is displayed to the user in order to obtain the MARR rate for the company from the decision-makers.

The DSSCIM then computes the results and displays them as shown in Figure 7.18. Then an informative message is displayed on a dialog box guides the decision-maker to continue evaluation.

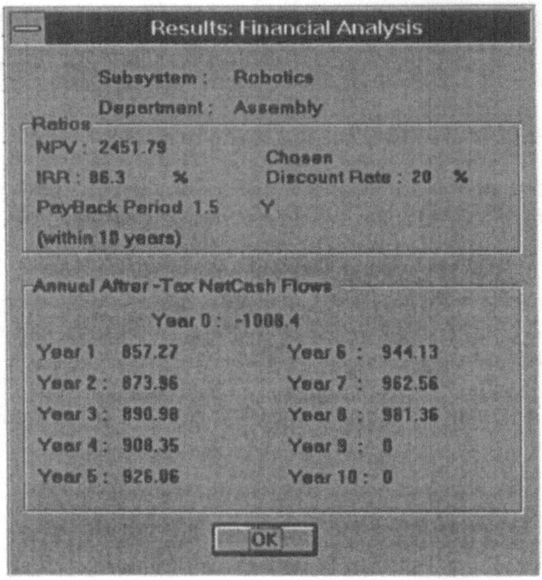

Figure 7.18 A dialog box displaying the financial outcome for investment in robots

From these estimations:
The total benefits which can be directly or indirectly measured
= \$981,900/for the Year 1
The total value of the attributes which are difficult to quantify
= \$50,000
Hence, for the typical example shown in Table 7.6 (page 118), the "Effective capital investment" = \$1238,000 – 123,800 – 50,000 = \$1064,200

The parameters for the calculations are considered as follows:

- Straight line depreciation method is used
- Salvage value is \$30,000
- Life of the asset is 8 years
- Tax rate for taxable income throughout the study period is 15%
- MARR for the computation of NPV is 20%

This exercise illustrates how the consideration of intangibles in financial evaluation can provide results which are both realistic and practical. Nevertheless, as explained in Chapter five, when the reduction of labour is considered only as a benefit obtained from the introduction of an AMT, the financial evaluations will not produce satisfactory results. For all other investments considered in this example, the financial evaluations are not given. However, the principles are the same as explained above. On completion of financial analysis for all AMTs, by selecting the "Analysis" option, the DSSCIM generates the results shown in Figures 7.19 and 7.20.

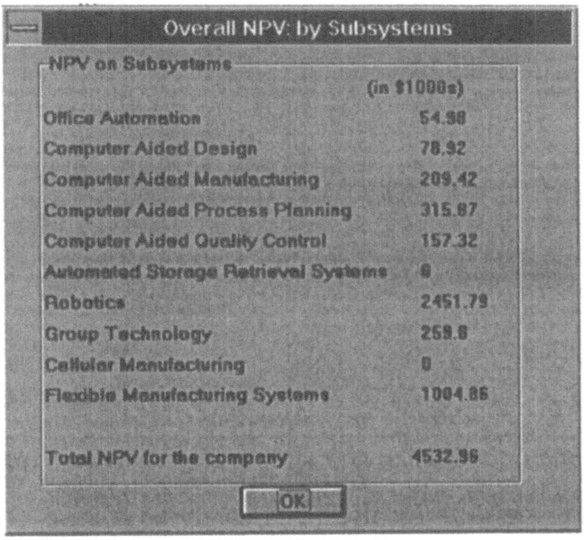

Figure 7.19 Financial results by subsystems

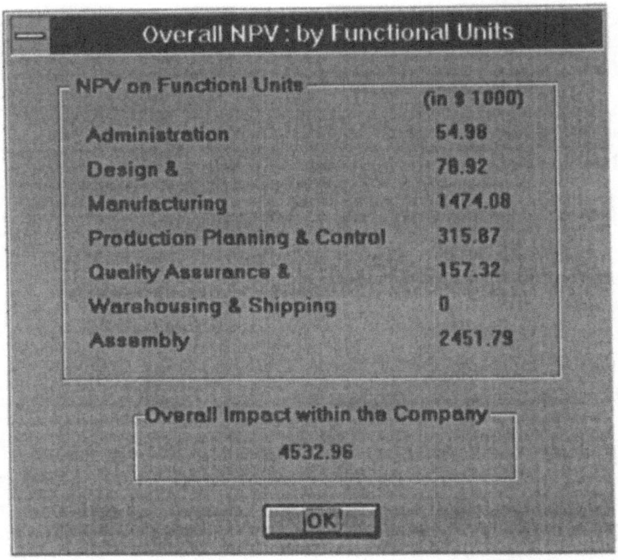

Figure 7.20 Financial results by functional units

Then the definitive cost-impacts illustrated in Chapter five (Section 5.2.2.2) are computed by the DSSCIM and displayed as shown in Figures 7.21 and 7.22.

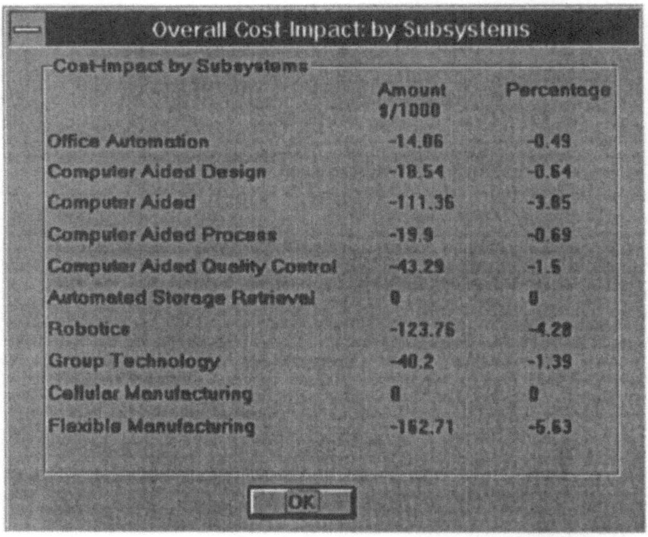

Figure 7.21 Definitive cost-impact by subsystems

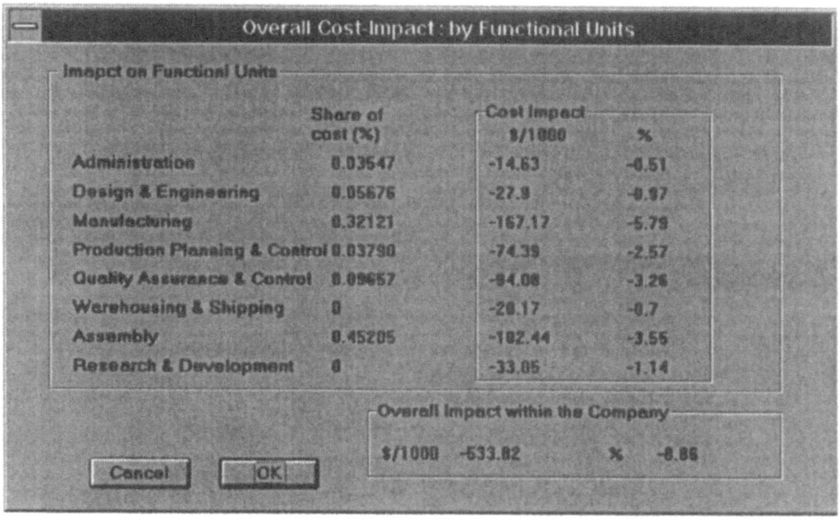

Figure 7.22 Definitive cost-impact by functional units

Then the DSSCIM displays a dialog box to determine the preference of the decision-maker in optimising their investment capital.

7.5.7 Optimisation

Once the selection is made for optimisation, the DSSCIM selects the optimisation worksheet and displays a dialog box about the succeeding dialog box. Then it displays the dialog box shown in Figure 7.23 (page 124) to assign a feasible minimum and maximum amount for the AMTs in conformity with Figure 4.1 (page 55). By contrast, if the decision-maker does not prefer the optimisation analysis, the operation of the DSSCIM is terminated and a printout, if requested, is given to the user for the analysis conducted to that point.

Once the allocation is made, the DSSCIM executes the GP algorithm and displays the results of optimal allocation for capital investment which satisfies the strategic objectives of the company.

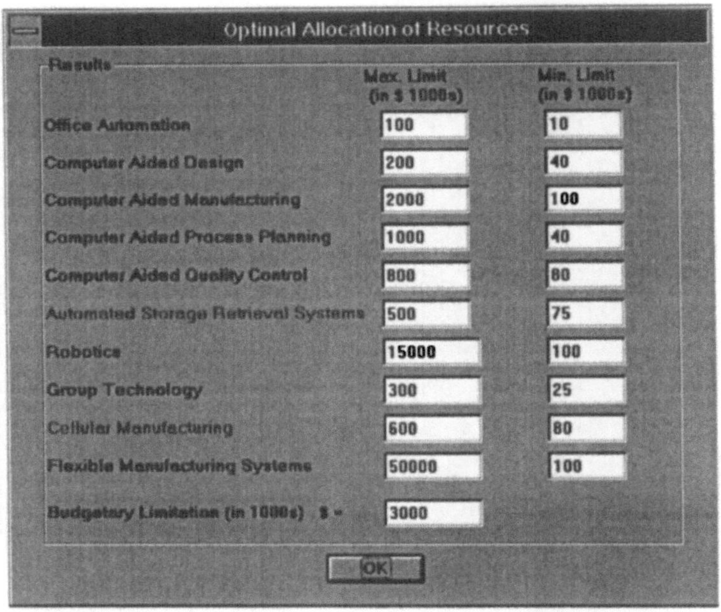

Figure 7.23 A dialog box for the allocation of capital limits on pre-selected AMTs

7.5.8 Results

The results obtained in this example are shown in Figure 7.24 (page 125). On the request of the user, the DSSCIM provides a printout of the results in hard copy.

This example clearly illustrates irrespective of the high hurdle rate of 20% considered in the financial evaluation, investments in AMTs are easily justifiable within the financial expectation of the company when the intangible benefits are incorporated into the justification procedure. The definitive cost-impacts on the functional units as shown in Figure 7.22 (page 123) illustrate the effect of synergy: even though investment in "Warehousing and shipping" and "Research and development" are not considered, those functional units also show significant improvements in the reduction of operational costs. In addition, the company will experience a reduction in operating costs of $533,820.0 or 8.86% in the first year of introducing AMTs. This is a significant improvement for the company. The optimal resource allocation shown in Figure 7.24 (page 125) indicates that having a large investment in FMS, with mostly minimum investment allocated to other AMTs, will help the company reach its strategic objectives within the available budget through the effect of synergy and integration.

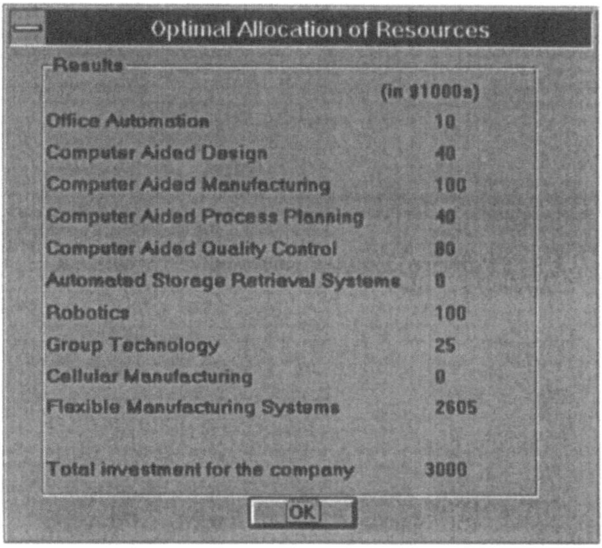

Figure 7.24 A dialog box displaying optimal capital allocation

On completion of the justification and optimisation of the AMTs, one of the significant tasks of the managers is to ensure that the expectations of the company are fulfilled by the implementation of the AMTs. In order to achieve a successful result from CIM implementation some necessary guidelines have to be followed, which are addressed in the next chapter.

7.6 CONCLUDING REMARKS

The example considered in this chapter illustrates the usefulness and capability of the proposed methodology and the DSSCIM to a manufacturing company. By utilising the DSSCIM, the company will be able to evaluate and estimate the cross-functional benefits and impacts derived from the investments in the integrated AMTs. The example demonstrates an application of the Cost-Value technique to estimate the value of an intangible benefit and to incorporate that into the evaluation procedure. Use of the Cost-Value technique and the mechanism of estimating savings from the database provide a realistic analysis to justify investments in CIM. The optimisation technique of the DSSCIM provides a facility to allocate the capital investments to satisfy the strategic requirements.

7.7 REFERENCE

1. Gerrity, T.P., 1971, "Design of Man-Machine Decision Systems; an application to portfolio management", *Sloan Management Review*, Winter Issue, pp. 59–75.

CHAPTER EIGHT

Guidelines for Implementing CIM

8.1 OVERVIEW

Can the manufacturing industry revitalise itself, once the decision-makers have identified a strategy for investing? If we help them realise the potential of the new technologies available today will they benefit from it? Why are the managers failing in implementing in AMTs? These are the questions we should address to understand and resolve the problems in utilising AMTs.

8.2 IMPLEMENTATION GUIDELINES

CIM is not a turnkey system; therefore standard implementation procedures cannot be bought anywhere. Inadequate organisational and operational planning for the adoption of AMTs, and incompatible procedures in the implementation process have resulted in many failures in adopting AMTs. The path to successful implementation of CIM is full of pitfalls. However, understanding the difficulties and the reasons for failures will prepare manufacturing managers to handle the implementation approach in a positive way. Many reasons for failures in CIM implementation have been reported in literature and can be summarised as:

1. failing to recognise the strategic benefits of CIM, and failing to view the principles of AMTs from a broad perspective;
2. failing to prepare the implementation with carefully measured steps;
3. expecting huge benefits and unrealistic productivity improvements in a short time;
4. over-estimating the average utilisation rate of the proposed new technology as a basis for appraising benefits;
5. improper planning and insufficient technical expertise in solving problems in an AMT environment;
6. failing to establish at an early stage, the architecture for compatible integration among various technologies;
7. under-estimating the magnitude of changes necessary to move towards CIM;
8. under-estimating the time needed to achieve effective functioning of the AMTs;
9. lacking communication standards to integrate islands of technologies;

10. lacking knowledge in the selection of appropriate AMTs to support a corporate plan;
11. lacking management awareness and commitment;
12. under-estimating the problems and costs of gaining labour acceptance for changes in tasks due to the introduction of AMTs;
13. failing to recognise the resistance to change at all levels in the organisation;
14. failing to simplify the implementation process;
15. failing to provide maintenance programs that focus on bottlenecks or having inadequate maintenance standards; and
16. lacking management awareness to make adaptive adjustments to the operations of an integrated production system, or to the re-allocation of support resources between the new facility and the older facilities.

Any company that embarks on AMT or CIM will necessarily experience unprecedented changes in the nature and scope of implementation. CIM will have an impact on all manufacturing activities in a radical manner and consequently will have a major impact on all aspects of organisation and management. Most importantly, commitment of senior management, a drive towards revitalising the company and careful planning is imperative for the success of AMTs. The implementation of CIM will necessitate changes in all areas including plant layout, data flow structures, production and design concepts, and accounting and cost functions. Hence, the following guidelines have been proposed for the success of the implementation of CIM/AMTs:

1. Since the impact of AMTs transcends traditional functional units, to derive success in CIM implementation, it is imperative to form cross-functional teams to evaluate the planning of CIM. Many management articles postulate that manufacturing managers play a pivotal role as product champions in the adoption of AMTs. Hence, a leader among manufacturing managers has to be identified and given the responsibility of implementing CIM with the help of the cross-functional teams.

2. Since CIM affects all areas of a company, every individual unit should participate in the development of the CIM strategy and its planning process. As many companies have discovered, it is essential that an employee identify himself or herself with the implementation of CIM for its success.

3. Since implementation of CIM requires restructuring of the labour force, appropriate education and training should be provided to the employees to obtain skills in managing and operating advanced technologies in a confident manner. This training and education process should also address the concern of the displaced employees and orientate them towards using the new technology. By providing new skills to the employees who have been relocated by the introduction of CIM from previously held positions, the morale of the employees will improve and support for CIM implementation can be easily achieved.

4. Since CIM has the potential to reduce the number of middle management activities, the management should also be given training on CIM and its benefits, thus overcoming their initial resistance in adopting new technologies.

5. Since participation of the supplier in the enhanced process is essential in providing more frequent deliveries to meet the increased demands, a very good supplier relationship should be established. At the same time, to compete effectively it is necessary to establish a good communication channel and positive relationships with customers in order to get direct feedback about the products.

Planned investment in CIM will generate improvements in virtually every element of a company. Factors such as: improvements in quality and data flow among various units; and reductions in the areas of inventory, material costs, lead times, labour costs and overheads, contribute towards lowering the unit costs of the product, thereby improving profitability and the ability to compete in a highly competitive market as well as generating opportunities for increased output. In order to achieve proper planning and successful CIM implementation, a flowchart for implementation can be used (see Figure 8.1, page 132). The flowchart represents eight stages of CIM implementation for clear understanding.

8.2.1 Implementation Stages

CIM is a radical investment. Hence, as we have emphasised earlier, identifying the strategy of the company is paramount for the success of the investment. CIM creates apprehension in the mind of many employees. Therefore, it is necessary to create a team consisting of all the departments or functional units of the company and empower them with implementing CIM. It has been observed in many companies, a champion for the cause and the participation and commitment of the top management is essential for the success of new technologies. The cross-functional team should have the blessing of the senior management to make appropriate changes in the working environment of the company.

As we have seen in Chapter four, identification of the initial components of CIM that will satisfy the strategic objective of the company is a crucial stage in implementing CIM or any new AMTs. Acquiring know-how on the components of CIM will eliminate many problems in later stages of implementing the new technologies. In addition, as we have seen in Chapter two, identifying technologies which are complimentary technologies to each other will assist in realising synergy among them. Similarly, determining technologies which are substitute technologies for other AMTs will eliminate duplication. Therefore the implementation team for CIM should delve deeply into these issues before proceeding further. Once implementation issues have been identified, the concept of CIM and the impending changes should be disseminated as widely as possible in the organisation.

Educating employees plays a critical role in reaching the success of CIM. Since, it ensures the fear of CIM is removed or reduced from the minds of the employees and it will help them to identify new skills and knowledge they can develop in order to participate fully in the CIM program. This employee-oriented approach, in revitalising the manufacturing strategy of the company acknowledges support from everyone, is important for the success for CIM and prepares them towards embracing it happily. In addition, CIM education programs will improve the morale of the employees.

Purchasing new technologies and evaluating the tendering process from various vendors is a time-consuming process for an organisation. In addition, reliability of the vendors plays a major role in obtaining on-going support for any future technical glitches or difficulties with the AMTs. Therefore, the prospective vendors should be identified and their capabilities should be assessed before any commitments are made. On the other hand, if the vendors are not able to provide on-going technical support, a cost-effective way of obtaining technical support for the implementation should be identified. This can be in the form of developing in-house know-how or out-sourcing to other technical support groups. In addition, establishing a co-operative approach with vendors will assist in overcoming many vendors and suppliers related problems in the future.

Financial justification is the main concern of any investment. The decision-makers, who are members of the cross functional team, should be able to justify the investments with the participation of the senior management. The decision-makers can utilise the concepts explained in Chapters four to six in evaluating AMTs and the associated benefits with tangible and intangible attributes of the AMT.

Once the justification is performed, optimising the resources within the strategic objective of the company is necessary to gain the benefit of the investment in an optimal manner throughout the company. The optimisation procedure proposed by the authors in Chapter six can be utilised by the decision-makers for this purpose. The optimisation procedure will enable selecting combination of AMTs among initially identified technologies and allocating them in appropriate functional units to work in an integrated manner to obtain the benefit for the whole company.

Once the components of CIM and the size and technical capabilities of them are identified for implementation, the decision-makers should focus their energies towards preparing the conceptual layouts of the integrated system. The layouts are then improved and the location of each component is verified in the enterprise. In addition, issues such as data sharing facilities, storage and retrieval areas of the products and the working environment for the equipment are to be resolved with the participation of all functional units concerned.

The final stage of the implementation process is to install the equipment and the software components associated with it and test run the integrated system. The performance of the system should be assessed and any difficulties in the testing stage should be resolved before commencing full-scale operation. However, even if bugs are found after activating the full production schedule, we will be able to rectify the problem and improve activities in a CIM system. Once the system has begun production, the skilled technical staff need to monitor the system for a while to identify any glitch which was not foreseen. Theoretically, glitches should not exist in a properly planned and implemented system. However, in practical reality it may pose a serious concern to the operation and the viability of the new system. It is advisable to prepare preventive maintenance programs for the CIM system to gain the full benefit of it. These preventive maintenance programs should be accommodated within the production scheduling arrangements.

The implementation procedure can be illustrated by the flowchart shown in Figure 8.1, page 132. The flowchart shown in Figure 8.1 is explained in following steps:

Step 1: Identify strategic requirements

- Establish a cross functional management team empowered to make necessary changes
- Break down the requirements into identifiable attributes
- Create an implementation task group
- Appoint a leader to champion the task
- Develop implementation strategies and plans

Step 2: Identify preliminary AMTs needed

- Carefully identify the AMTs which can provide the strategic requirements of the company
- Identify all tangibles and intangibles for each AMT
- Prioritise the importance of the attributes and assign weighting scores using a structured approach such as AHP
- Identify any combinations of the AMTs which can satisfy strategic requirements
- Identify the benefits of integration and synergy

Step 3: Inform and educate employees

- Inform employees about the introduction of AMTs
- Educate all employees to understand the benefits of AMTs
- Involve everyone in the company
- Resolve any confusion and conflicts
- Identify necessary new skills
- Overcome any resistance to change by re-training
- Identify and prepare learning facilities for key personnel
- Plan continuous training programs for all employees

Step 4: Contact prospective vendors

- Identify reliable vendors
- Evaluate their capacity to provide maintenance and ongoing support
- Request preliminary quotations
- Check compatibility of equipment from different vendors

Step 5: Perform financial justification

- Evaluate benefits and values of intangibles
- Compute financial indices while incorporating intangibles
- Compute impact on other functional units due to synergy

Step 6: Determine the technological level of AMTs

- Identify limitations on resources
- Utilise optimisation techniques
- Determine a combination of technology and a budget limitation for the strategy

Step 7: Prepare implementation

- Prepare conceptual design and layout for systems within the enterprise
- Prepare hardware and software acceptance tests
- Evaluate the systems to be used
- Prepare a detailed layout and plan for subsystems including integration architectures

Step 8: Install and integrate system

- Install equipment
- Install software and integrate through information networks
- Test the system for functionality and evaluate performance
- Train employees and evaluate their performance
- Prepare maintenance procedures for software and hardware

8.3 CONCLUDING REMARKS

Many articles suggest implementation of AMT, based on the authors' experience and literature survey. These suggestions give varieties of methodologies to manufacturing managers and do not give an approach which guides the managers through the implementation process. Many papers merely suggest what to do for the implementation of AMT and do not express how to perform the justification process. The authors of this book believe that the literature should not limit itself only to providing suggestions on how to implement AMT or what to do with an AMT investment. The literature should also provide mechanisms and decision support aids to improve the manufacturing environment by investing in AMTs. Keeping that concern in mind, a methodology to invest in CIM (see Chapter four to six), a decision support system for justification and optimisation (see Chapter seven), and in this chapter the guidelines to implement the system have been provided.

The intention of the guidelines proposed in this chapter are to help manufacturing managers understand the necessary procedures involved in the adoption of AMTs or any integrated system such as CIM. "If we fail to plan, we are planning to fail" is a widely known adage. By adhering to the guidelines, the companies will be able to plan the implementation of AMTs in an appropriate manner and avoid costly mistakes.

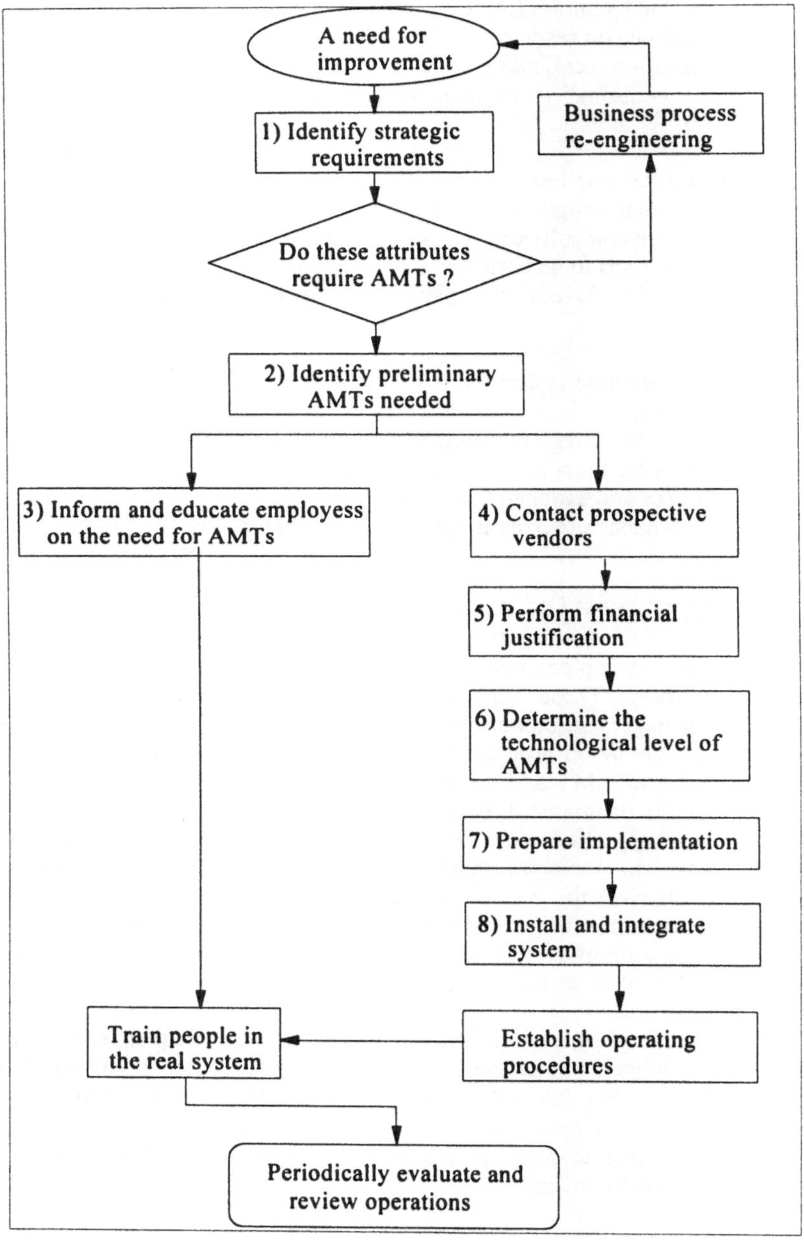

Figure 8.1 Flowchart for implementation guidelines

CHAPTER NINE

Conclusion

A man who does not plan long ahead will find trouble right at his door.
Confucius

9.1 CONCLUDING REMARKS

The purpose of this book is to make manufacturing managers become aware of the changing needs of the manufacturing environment and equip them with a valuable tool for justifying and optimising the investments in AMT and CIM.

The global marketplace is becoming increasingly competitive. Hence, manufacturing organisations are taking profound interest in seeking new methods to optimise efficiency and productivity. To fulfil these criteria, the organisations should be proactive, flexible and adaptive to changes. Nevertheless, manufacturing has entered an era of drastic and improved development with the help of AMTs. Increased automation and flexibility provided by AMTs are the main propelling forces behind implementing the factory of the future, where all the operations related to a manufacturing company are accomplished with an extremely low level of human interaction. Companies which are reluctant to upgrade their technologies to AMTs will have tremendous difficulties in competing in the dynamic market.

CIM as an emerging technology, brings new perspectives to the decision-making and its justification. Moreover, progressive changes are required in the management approaches to deal with CIM. Justification of CIM requires a strategic approach, analytical evaluation techniques and an economic analysis. A decision to implement CIM represents a major capital investment and commitment from the company. A strategic approach is an important aspect of any goal oriented enterprise. Appropriate analytical techniques are mandatory in providing structured approaches to decision-making. Economic justification of any new investment is paramount for survival of the commercial enterprises, and any new investment should provide satisfactory outcomes, at the same time, to both customers and shareholders.

AMT has increased the share of indirect costs and reduced direct labour costs from the investment analysis. Literature abounds with criticism of the traditional justification techniques. Traditional techniques do not have the capability of justifying the investments in innovative technologies such as CIM. Furthermore, the traditional capital budgeting techniques favour short-term returns and ignore strategic considerations and intangible benefits/costs of an investment in AMT. This is a consequence of the inadequacy in incorporating strategic issues and quantification of the intangibles by the traditional justification techniques. Conventional techniques are only appropriate for either equipment replacement decisions or decisions to implement are to be made based on reduction in direct labour. In addition, manufacturing managers will not be able to incorporate the qualitative factors, which could enhance the efficiency and the competitive position of a company, by relying on these mundane justification techniques. The incorporation of qualitative measures in a justification process with other pertaining issues also included into an integrated approach is not well established in any previous research. Hence, research was conducted at the Centre for Advanced Manufacturing Research, University of South Australia, to provide an innovative and appropriate approach as a unified justification methodology for the decision-makers of companies.

CIM is a management and integrated manufacturing approach, which allows companies to manufacture products with flexibility at reduced cost. In addition, it allows the companies to provide swift responses with minimum resources. However, the application of the CIM requires an in-depth, but a simple evaluation approach to incorporate all the required issues pertaining to an investment. The research conducted at the Centre for Advanced Manufacturing Research is a relevant step in that direction which provides a simple but an effective methodology, and a DSSCIM for manufacturing managers to analyse the benefits of the investments in an integrated environment. The developed DSSCIM provides a multi-attribute decision-making mechanism that accounts for both quantitative and qualitative factors that affect an investment decision.

Managers generally feel confident in making decisions that involve certainty and predictability. Hence, it is imperative to provide a DSS which has the capability to assist manufacturing managers to reach a decision in a confident and viable manner. By using the DSSCIM developed in this research, in addition to a comprehensive evaluation of AMTs, a decision-maker can achieve:

1. an understanding of the objectives for the CIM implementation and the methodologies required in justifying AMTs in a CIM environment;
2. an understanding of possible AMTs which are related to the strategies in pursuance by the company; and
3. an objective decision on the expenditure of the investment by recognising the potential benefits, synergy effects, impacts on the individual functional units (and on the company as a whole), and optimised capital allocation while satisfying the strategic objectives.

Success of a commercial company depends on the financial performance and is measured by the contribution of the investment towards the profitability of the business. The difficult task facing a decision-maker is to combine the modern economic analysis and the subjective considerations in the strategic analysis in a constructive manner to reach an optimal decision. In the customers' perception, the financial consideration is not the only criterion for making a decision when purchasing a product. A variety of non-monetary attributes and multiple objectives such as quality, appearance, utility, durability and value are considered by prospective customers. Therefore, when evaluating the investment decisions for product facilities, these multiple criteria should also be addressed. Some of the most common non-monetary attributes that must be considered are: general business conditions; social and human values; consumer preferences; environmental concerns; and government regulations.

Evaluation of advanced manufacturing systems requires a multi-attribute decision environment which could analyse impacts of the investment in an all encompassing manner. The provision of a strategic decision profile to identify and suggest the suitable AMTs which satisfy the strategic requirements is one of the key features of the proposed methodology. Incorporating the multiple attributes and evaluating them by ranking, imposes subjectivity in the decision-making process. The proposed methodology, with the help of the AHP, reduces subjectivity in the analysis to arrive at a final decision as whether or not to invest in CIM, by providing structured approaches in evaluating the subjective parameters in the decision-making exercise. However, a weighted scoring approach is also included for prioritising the attributes for a manufacturing manager who is appropriately experienced in evaluating the multiple attributes.

Planning for a CIM environment requires a detailed evaluation of dynamic interactions among various subsystems. Incorporation of a functional inter-relationship matrix and the provision of a framework to evaluate the synergy among integrated AMTs is another key feature in the proposed methodology. Awareness of possible interactions among different subsystems of the CIM, as synergy effects, helps in evaluating the company-wide impact of new technologies in a realistic manner. In addition, as a simple feature, the impact is evaluated by identifying the benefits of each attribute on the AMT in the prime functional unit only. Then the DSSCIM allows the computation of the company-wide benefit by taking into account the identified synergy. Identifying the impacts of each AMT in the CIM framework provides a basis to incorporate all the features of AMTs in the evaluation procedure and the appreciation of the new technologies.

The use of a database in assisting the estimation of the value and benefits of an attribute is an important feature towards reducing the knowledge gap in the justification process; thus helping the manufacturing managers to make investment decisions on AMT. Value judgements are made possible with previous knowledge and past experience gathered from the literature, thereby reducing the subjectivity of the decision. While the Cost-Value technique provides an effective approach in dealing with the intangible aspects in AMT justification. In addition, disintegration of intangibles into tangible attributes provides an additional mechanism to quantify many intangibles previously considered not quantifiable.

The goal programming technique provides a realistic approach to optimise manufacturing resources while satisfying multicriteria requirements. The proposed methodology further strengthens this optimisation approach by incorporating the strategic priorities identified in the strategic decision profile. In particular, by providing a unified approach in identifying strategy and suitable subsystems, evaluating departmental and company-wide impacts, quantifying intangibles for financial justification, and incorporating synergy in impact evaluation and in optimisation of resources, the proposed methodology emerges as a significant justification and optimisation technique for CIM investment analysis compared to existing methodologies.

The investment analysis methodology proposed in this research does not advocate discarding current accounting practices; rather it provides an efficient and effective methodology to incorporate the intangibles and synergy in evaluating the investment as an internal assessment tool. The current accounting practices can still be used as external reporting tools to satisfy the auditors and shareholders.

However, by educating the external participants on the advantage of an effective method, it is expected that they will also embrace the positive approach provided by this methodology.

The introduction of CIM or AMT in an existing company should carefully address the psychological implications on all employees in the enterprise. Naturally, automation and the introduction of advanced technologies generates employees' apprehensions of being made redundant and incapable for further employment with existing skills. Such concerns generate impediments and unhelpful attitudes towards adopting new technologies. The authors firmly believe, if any existing company should adopt a new concept such as CIM, it should introduce it in a planned manner: throughout the period of change all the employees including shop floor workers, middle managers and senior managers, should be given appropriate training and counselling to increase their acceptability towards newer technologies. These concerns, however, do not affect a new company which is going to be established as a green-field company. Therefore, managers of the green-field company can proceed with full vigour in the introduction of new technologies, thus gaining a superior competitive edge in the global market.

9.2 THE SIGNIFICANCE OF THE APPROACH

The significance of the proposed methodology is to provide a unified approach with many features which was not attempted earlier. In particular, the functional inter-relationship matrix, and the use of the Cost-Value technique to quantify intangibles, besides utilising the quantified intangibles for the financial evaluation, contributes an innovative direction in the justification of CIM. Furthermore, the methodology proposed for objectively evaluating the subjective indices of the functional inter-relationships provides a notable contribution towards evaluation and greater understanding of enterprise integration. By incorporating synergy in impact evaluation and optimisation, a new worthwhile dimension to the justification is included. Selection of suitable technologies and optimised resource allocation, while satisfying strategic requirements enhances the usefulness of the DSSCIM for strategic decision-making.

The DSSCIM developed as a prototype, which is explained in Chapter seven, guides the decision-maker as an interactive computer program to make the best possible combinations of the solutions under the circumstances. The DSSCIM provides a structured approach in reaching an investment decision for an integrated enterprise and an ideal DSS for AMT evaluations using a multicriteria approach. Managers and decision-makers can determine their investment decisions within their company's own comprehensive set of criteria with the help of the DSSCIM. In addition, the incorporation of the quantification of intangibles eliminates a significant hurdle in AMT justification. These procedures in the DSSCIM assist manufacturing managers to realise the potential of incorporating such intangibles into the justification procedure. The attention of the decision-makers is guided towards evaluating synergy among AMTs by the use of a functional inter-relationship matrix. By reminding managers and decision-makers to develop a data-flow diagram across the functional units and AMTs, the DSSCIM further guides the decision-makers to consider the simplification of the data transfer. The main characteristics of the developed DSSCIM can be described as shown in Figure 9.1.

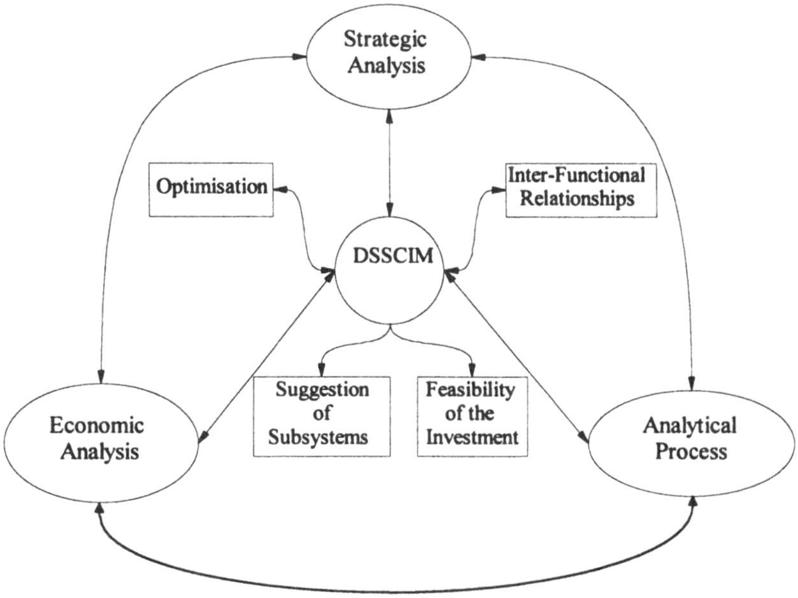

Figure 9.1 Characteristics of the DSSCIM

The important features of the proposed approach and the DSSCIM compared to other currently available multi-attribute analysis methodologies or decision support systems (DSSs) are listed as follows:

- It provides a realistic and unified approach;
- It focuses on identifying the strategy and provides a practical approach in relating the identified strategy in the selection of AMTs;
- It provides a simple and a guided approach for manufacturing managers with interactive dialogue frames;
- It provides a facility in realising the benefits of AMTs;
- It combines the strategic issues, and analyses with AHP in detail, if that option is preferred by the decision-makers;
- It provides the flexibility of simplifying analysis by having a number of user-driven options;
- It is effective in quantifying intangibles and indirect costs;
- It provides an effective approach for incorporating synergy;
- It is more user friendly by the provision of a database in estimating the benefits; and

- It provides an optimisation approach that satisfies requirements of a multicriteria decision-making environment.

9.3 FUTURE DEVELOPMENTS OF CIM

Research provides a basis to improve our existing knowledge and capabilities. Everywhere, our knowledge is limited and many problems need to be solved. Therefore, a necessity exists to address the gaps in knowledge and search for solutions for the unresolved problems. The role of research is to provide a method for obtaining answers to the problems by analysing the facts within the scientific method. However, a research process never reaches finality; the end of a research process is the beginning of another. Bearing this in mind, some suggestions are given in this section as possible future developments, which would further the application of CIM in the evolving and dynamic manufacturing environment. Predicting the future research direction of CIM and related areas is a difficult task in an ever expanding and growing technological development era. However, an attempt is made to foresee the future direction, which will dominate the researchers' mind for the next decade, based on the current developments in CIM research.

The primary challenge of being successful in a market depended on price based competition in the early 60s, and in the early 80s it evolved to quality and price. Today, the time taken to market a product is the third competitive factor. In addition, to meet the present day challenges of global competition, as well as to satisfy the expectations of the future, manufacturing companies should be dynamic, adaptive to changes and optimise their resources. The concept of virtual manufacturing where the functional units of manufacturing conglomerates are globally distributed in order to meet the resource optimisation is an evolving feature, to be addressed in the present and future situations.

Manufacturing industries have evolved tremendously from the cottage industries in the early sixteenth century into a global force as they are today. In today's competitive global market survival of any industry depends on its ability to communicate and transfer the right information at the right time to the right people. Manufacturing cannot escape from this present requirement. Having an ability to communicate for effective management and manufacturing activities across the geographical boundaries among globally distributed resources will significantly benefit manufacturing industries. Today, a number of global conglomerates are formed in many facets of industry. A virtual enterprise is defined as a network of interconnected global conglomerates in this chapter.

Today's competitive and agility requirements of the global market can only be met by virtual enterprises. To provide a better future in the present market requirements, research in virtual CIM and the application of it in world-wide manufacturing industries is beginning to emerge. Application of virtual CIM has been proposed as a necessary step towards the future in manufacturing to face competitive challenges (Lin, 1997). However, many development works need to be carried out to face challenges faced by virtual enterprises (Makatsoris and Besant, 1997). Hence, the research should be further strengthened towards developing a virtual CIM to satisfy the globalised and distributed manufacturing enterprises of today in order to meet the competitive and agility requirements of present market conditions.

In a virtual enterprise the integration of information is extolled, as only through information can a virtual organisation become meaningful, and only by effecting a new generation of information technology can this vision be realised (Pant, *et al.*, 1994). To represent the evolving process of CIM and to reflect the present need for a Virtual CIM, which stresses the importance in strategic and integrated management in implementing CIM in globally distributed enterprises, a new CIM wheel (Virtual CIM Wheel), as shown in Figure 9.2, has been developed at the Centre for Advanced Manufacturing Research, University of South Australia.

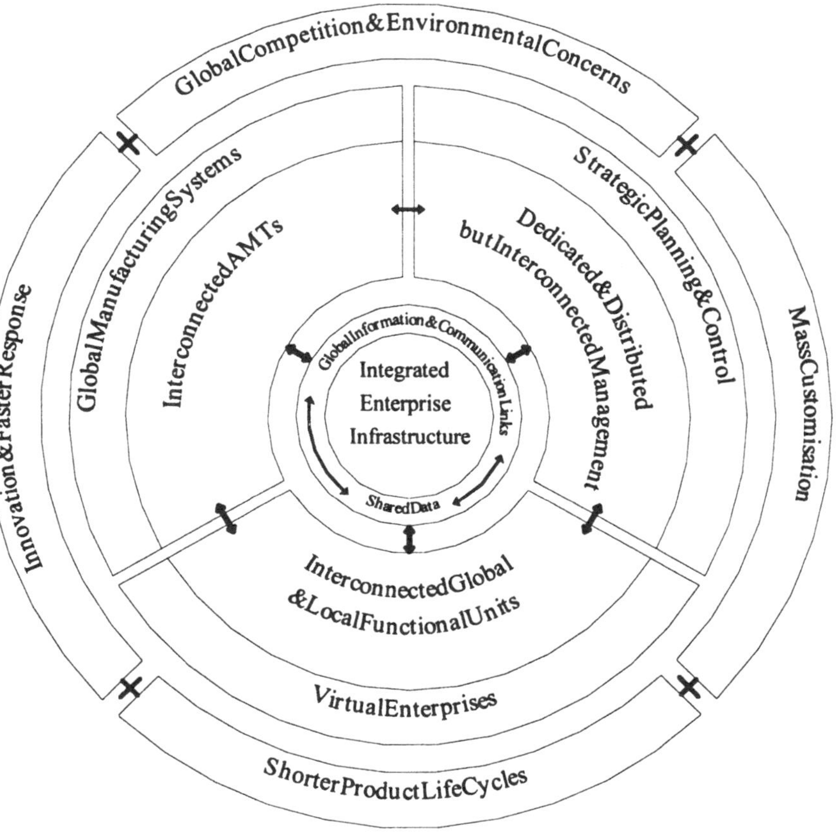

Figure 9.2 Virtual CIM Wheel to represent the global market conditions

The Concept of the Virtual CIM Wheel is in parallel with the CIM wheel developed by the Society of Manufacturing Engineers and it can be explained as follows (The explanation is given from the outer layer as the starting point):

- The outer circle represents the present world situation. It depicts characteristics such as Global competition; Environmental Concerns; Mass customisation to satisfy the variety of customer requirements; Shorter product life cycles of the product; and Requirement for innovative products and of faster response;
- The second circle from the outer layer represents global systems and concepts needed to address the situation;
- The third circle from the outer layer explains briefly how the concepts and systems can be realised;
- The fourth circle from the outer layer represents the need for global information and communication links and the need to share data among systems; and
- The inner circle represents the outcome of CIM as a globally integrated enterprise through an integrated architecture.

To facilitate the investment in CIM and to help manufacturing industries, many researchers are seeking solutions in the application of CIM. Some of the research trends in CIM which are evolving today can be categorised as follows:

- Justification of CIM and management strategies for CIM;
- Enterprise integration for CIM beyond and within geographical boundaries;
- Network communications for the implementation of CIM;
- Advanced tools and technologies for the application of CIM;
- Manufacturing system modelling; and
- Application of artificial intelligence (AI) such as fuzzy logic, neural network, genetic algorithms and intelligent agents for fully integrated intelligent manufacturing systems.

Each of this categorisation is diverse in nature and consists of many sub-groupings. In addition, few research directions may overlap in more than one category due to the complexity of the research in CIM.

Justification and management strategies for CIM consist of a multitude of research directions. This research trend focuses on providing managers with principles and guidelines for the application of CIM in their working environment and approaches to overcome the resistance to change towards integrated manufacturing. Further, this categorisation addresses the organisational and social issues in an enterprise, issues involving bench marking the features of the manufacturing supply chain and manufacturing strategy implementation through real-time systems, optimised enterprise integration and DSSs and expert systems (ES) for manufacturing managers. DSSs and ES may further include ES for quality inspection, ES for production planning and selecting manufacturing parameters, DSS to evaluate designing decisions, financial decisions, and educational and informative systems.

Enterprise integration for CIM beyond and within geographical boundaries consists of a research diversity, which includes architectures and modelling formalisms for enterprise integration, evaluation methodologies for enterprise integration, and international CAD/CAM collaboration for CE and CIM implementation through integration of subsystems. While network communications for the implementation of CIM category includes application of wide-area networks and Internet for CIM, information enhancement by data integration, issues related to integration of client and server for manufacturing shop-floor automation, application of multimedia and hypermedia in CIM environment, and data management in CIM systems.

The categorisation of advanced tools and technologies for the application of CIM is one of the thriving areas in CIM-related research. It consist of robotics and automation for a CIM framework, vision-based manufacturing systems and intelligent AGVS and intelligent and knowledge-based enabling systems for CIM. The manufacturing system modelling category consists of research variations which include integration of information models with functional models for CIM, object oriented resource modelling, integrated simulation modelling approaches for CIM, and modelling approaches and methodologies for designing CIM systems.

Application of AI for fully integrated intelligent manufacturing systems consists of research directions, which include the application of neural networks on manufacturing automation and technologies, intelligent scheduling systems with genetic algorithms, fuzzy scheduling systems, hybrid systems of CIM components, adaptive models for CIM architectures and embedding AI in CIM software.

9.3.1 Developments in Research Directions

In the research area that focuses on justification of CIM and management strategies for CIM, recent developments in the group-based management information systems have strengthened the application of a DSS in a group decision environment. In addition, rapidly changing technology and increasing reliance on knowledge and communication creates significant challenges to static and old models of an enterprise. Therefore, in order to justify and optimise investments in a virtual enterprise, a need exists for a Group Decision Support System (GDSS) which has the capacity to consider the effect of AMTs in a global framework. A GDSS can improve decision outcomes by using structured approaches for the unstructured problems in a collaborative and a distributed team oriented decision-making environment. A GDSS will help decision-makers in such a team environment analyse the investment decisions of a company in an interactive manner, either in a geographically distributed enterprise or a localised manufacturing enterprise with diverse functional units. In a localised manufacturing enterprise which has diverse functional units the use of GDSS can help the enterprise as follows:

- Senior managers could help in identifying the strategic goals of the company and prioritising them;
- Middle level managers will be useful in assessing the inter-relationships between AMTs and various functional units of the company;
- Knowledge inputs from engineering managers will be useful in establishing the performance measure for each criterion;

- Financial analysts with the help of middle level managers could input financial data for the investments and quantify the intangibles.

The rapid developments in technology provide innovative approaches to manufacturing and the recent trend towards global manufacturing have increased the number of geographically distributed and multi-location enterprises. In the research direction of enterprise integration for CIM beyond and within geographical boundaries, a collaborative framework for remote machining (Kao and Lin, 1997) and planning and control systems (Makatsoris and Besant, 1997) has been proposed. The prime technology which propels the advancement in manufacturing systems and international networking is the communication technology (Rembold, *et al.*, 1994). Hence, the research direction of network communications for the implementation of CIM has produced a plethora of research articles. In addition, the use of Internet and the World Wide Web in manufacturing applications is gaining momentum. However, research should be further strengthened towards providing integrated architectures and communication protocols for the virtual CIM framework to enable real time machining, processing, and decision-making facilities in geographically dispersed enterprises.

Research in robotics and automatic guided vehicles is predominant in the research direction of advanced tools and technologies for the application of CIM. In today's technologies the remote manipulation of these machines has already been achieved. Mobile robots and global positioning satellites have played a significant role in remote manipulation of robots (Feng and Kubik, 1996). Collaborative robots are being developed to machine complex surfaces and large components (Lin, *et al.*, 1998). In addition, vision-based systems have been developed towards rapid prototyping, part identification and processing. These developments need to be enhanced towards providing practical solutions for a globally dispersed manufacturing enterprise and should be tested in a virtual CIM framework.

Today, many manufacturing industries are under pressure to achieve their objectives within limited capital available for new investments. Optimisation of resource allocation for strategic investments could provide a mechanism to achieve the goal of the enterprise. As Hess (1992) has stated, the future of CIM depends on optimised integrated systems. Traditionally, to improve the effectiveness and competitiveness of an organisation, all the efforts are focused on achieving a single goal or achieving the objective within a functional unit. However, the traditional method of a single goal seeking mechanism or optimising the results of an individual unit, does not satisfy the overall objective of a manufacturing company. Therefore, a multiple goal seeking or multicriteria optimisation which considers all the relevant factors and overall benefits of a company as a single unit, has been proposed to make CIM the effective application in the present global economic circumstances (Nagalingam and Lin, 1997, 1998). The existing multicriteria optimisation mechanisms should be further enhanced or new approaches developed for the application in globally distributed virtual CIM enterprises, so that the benefits of virtual CIM are fully realised. In addition, global integration approaches which regards a globally distributed company as a single entity, should be developed and utilised to make virtual CIM the effective application to meet the present global economic circumstances. Intelligence Manufacturing is the path for the future. Therefore, manufacturing technologies should be embedded with intelligence, which will help manufacturing enterprises to produce better quality products and make manufacturing equipment to solve problems encountered during the normal course of operation itself.

9.4 REFERENCES

1. Feng, Y. and Kubik, K., 1996, "Real time GPS dynamic positioning approaches to mobile robot navigation", *International Symposium on Signal Processing and its Applications*, ISSPA, Australia, pp. 258–261.
2. Hess, J.G., 1992, "Beyond CIM-optimised enterprise integration", *Proceedings of Autofact '92 Conference*, USA, pp. 2–9 to 2–24.
3. Kao, Y. and Lin, G.C.1., 1997, "Development of a remote CAD/CAM collaboration framework", The 1st *International Conference on Engineering Design and Automation,* Thailand, pp. 667–670.
4. Lin, G.C.I., 1997, "The latest research trends in CIM, Keynote paper", *The 4th International Conference on Computer Integrated Manufacturing,* Singapore, pp. 26–33.
5. Lin, G.C.I., Lu, T. and Yin, Y., 1998, "The development of a dual robot system for prototype machining", *Fifth International Conference on Automation Technology and 1998 International Conference on Production Research,* Taiwan, CD Rom.
6. Makatsoris, C. and Besant, C.B., 1997, "Semiconductor virtual enterprises: requirements and solution proposals", Keynote paper, The 4th *International Conference on Computer Integrated Manufacturing*, Singapore, pp. 13–25.
7. Nagalingam, S.V. and Lin, G.C.I., 1997, "A unified approach towards CIM justification", *Computer Integrated Manufacturing Systems*, Vol. 10 No. 2, pp. 133–145.
8. Nagalingam, S.V. and Lin, G.C.I., 1998, "A methodology to select optimal system components for computer integrated manufacturing by evaluating synergy", *Computer Integrated Manufacturing Systems*, Vol. 11 No. 3, pp. 217–228.
9. Pant, S., Rattner, L. and Hsu, C., 1994, "Manufacturing information integration using a reference model", *International Journal of Operations and Productions Management*, Vol. 14 No. 3, pp. 52–72.
10. Rembold, U., Lueth. T. and Ogaswara, T., 1994, "Intelligent manufacturing components for future CIM systems", *A post print volume from the IFAC (International Federation of Automatic Control) workshop*, Austria, pp. 1–10.

Appendix 1

A 1.1 THE COMPUTATION OF CONSISTENCY RATIO

Consistency denotes that when a basic amount of raw data is available other data can be logically deduced from it (Saaty, 1980). The consistency ratio (CR) of a pair-wise comparison matrix is a mathematical indicator for the consistency of the pair-wise comparison process.

The comparison matrix (Table 7.1 page 104) is utilised to illustrate the computation of the consistency ratio.

$$[A] = \begin{bmatrix} 1.00 & 3.00 & 3.00 & 5.00 & 0.33 \\ 0.33 & 1.00 & 0.33 & 1.00 & 0.20 \\ 0.33 & 3.00 & 1.00 & 3.00 & 0.33 \\ 0.20 & 1.00 & 0.33 & 1.00 & 0.20 \\ 3.00 & 5.00 & 3.00 & 5.00 & 1.00 \end{bmatrix}$$

As explained in Chapter three (page 43), the consistency of the values were calculated using the principal eigenvalue of the matrix (λ_{max}). The approximate value of the λ_{max} obtained as follows:

1. Multiply the matrix of comparison [A] on the right by the priority weight vector matrix [B] on the left. This will result in a new column vector [C];
2. Divide each component on the vector [C] by the corresponding component in the priority weight vector [B]. This will result in a new vector [D]; and
3. Divide the sum of the components in the vector [D] by the number of components in the vector [D]. The resultant value is approximately equal to the λ_{max}.

The priority weights for the matrix [B] are computed according to the explanation given in Chapter three (page 42). In mathematical terms the matrix [B] is called principal eigenvector. The eigenvector is computed by the third method among those given in Chapter three.

Initially, normalisation of the column is carried out by dividing the values in each column by the sum of that column. Then the elements in each resulting row is added and that value is divided by the number of elements in the row, thus achieving the process of averaging over the normalised column.

The sum of the columns of the matrix [A] are {4.87, 13.00, 7.67, 15.00, 2.07}.

The division of the matrix [A] by the sum of the respective column values yields, a normalised matrix of [A].

$$\text{Normalised Matrix of } [A] = \begin{bmatrix} 0.21 & 0.23 & 0.39 & 0.33 & 0.16 \\ 0.07 & 0.08 & 0.04 & 0.07 & 0.10 \\ 0.07 & 0.23 & 0.13 & 0.20 & 0.16 \\ 0.04 & 0.08 & 0.04 & 0.07 & 0.10 \\ 0.62 & 0\,38 & 0.39 & 0.33 & 0.48 \end{bmatrix}$$

The summations of the row values of the normalised matrix are:

$$[A^{*}] = \begin{bmatrix} 1.322 \\ 0.352 \\ 0.791 \\ 0.325 \\ 2.210 \end{bmatrix}$$

There are five elements in the row of matrix [A]. Thus dividing the row values of [A*] by 5 yields the column matrix of [B].

$$[B] = \begin{bmatrix} 0.264 \\ 0.070 \\ 0.158 \\ 0.065 \\ 0.442 \end{bmatrix}$$

The matrix C is obtained by the vector multiplication of matrices A and B.

$$[C] = [A] * [B] = \begin{bmatrix} 1.423 \\ 0.365 \\ 0.800 \\ 0.329 \\ 2.387 \end{bmatrix}$$

Matrix [D] is obtained by dividing each component on the vector [C] by the corresponding component in the priority weight vector [B].

$$[D] = \begin{bmatrix} 5.380 \\ 5.180 \\ 5.056 \\ 5.068 \\ 5.400 \end{bmatrix}$$

The approximate value of maximum eigenvalue λ_{max} is obtained by dividing the sum of the components in the vector [D] by the number of components in the vector [D].

Summation of vector [D] = 26.084 and the number of components in the vector [D] = 5.

$$\text{Thus } \lambda_{max} = \frac{26.084}{5} = 5.217$$

The deviation from the consistency is called a consistency index (C.I), and it is represented by $(\lambda_{max} - n)/(n - 1)$, where "n" represents the order of the square matrix.

$$C.I = \frac{\lambda_{max} - n}{n-1} = 0.054$$

The consistency indices were randomly generated for reciprocal matrices with reciprocals forced to the 1 to 9 scale and of the order 1 to 15 using the sample sizes of 100 and 500 [Saaty, 1980]. These generated values are labelled as a random index (R.I) and given for matrix of order 1 to 15. Table A1.1 shows the order of the matrices and their respective R.I values.

The ratio of C.I to the R.I for the matrix of same order is called a consistency ratio. The order of the matrix considered in this example is 5; therefore the R.I value is 1.12.

Then,

$$CR = \frac{C.I}{R.I} = \frac{0.054}{1.12} = 0.048$$

The pair-wise comparison is considered satisfactory if the CR of the comparison falls below or equal to 0.10. In this exercise the CR = 0.048, which is a satisfactory value.

Table A1.1 Order of a matrix and its R.I value

N (Order of a matrix)	R.I
1	0.00
2	0.00
3	0.58
4	0.90
5	1.12
6	1.24
7	1.32
8	1.41
9	1.45
10	1.49
11	1.51
12	1.48
13	1.56
14	1.57
15	1.59

A 1.2 PRIORITY WEIGHT AND CONSISTENCY CALCULATION FOR SECOND ORDER MATRIX

For the matrix of second order, the CR calculation for the AHP can be simplified as follows:

$$[A] = \begin{bmatrix} 1 & x \\ 1/x & 1 \end{bmatrix}$$

$$\text{Normalised Matrix of } [A] = \begin{bmatrix} x/(x+1) & x/(x+1) \\ 1/(x+1) & 1/(x+1) \end{bmatrix}$$

The summations of the row values of the normalised matrix are:

$$[A^*]= \begin{bmatrix} 2x/(x+1) \\ 2/(x+1) \end{bmatrix}$$

There are two elements in the column of [A]. Thus dividing the row values of [A*] by 2 yields the column matrix of [B], which is the priority weight matrix.

$$[B] = \begin{bmatrix} x/(x+1) \\ 1/(x+1) \end{bmatrix}$$

Thus for a second order matrix (consisting of two attributes), if the pair-wise comparison is x, then the priority weights for the attributes are x/(x+1) and 1/(x+1). For any positive value of x, the consistency of the matrix will be always satisfactory.

A 1.3 REFERENCE

1. Saaty, L.T., 1980, "*The Analytic Hierarchy Process: Planning, Priority Setting, Resource Allocation*, (New York: McGraw-Hill International).

Appendix 2

A 2.1 BENEFITS GAINED BY THE INTRODUCTION OF AMTS

For AMTs considered in the DSSCIM, the reported benefits, which were gathered from the literature, for the technologies, and considered in the analysis, which was proposed in this book, are listed below:

Benefits by Office Automation

Office Automation:

1) Offers an ability to process more business information
2) Provides faster turnaround of business documents and transactions
3) Reduces physical space
4) Reduces errors
5) Improves customer services
6) Offers better decision and decision-making capabilities
7) Reduces number of office staff
8) Improves the performance of the staff

Benefits by CAD

CAD:

1) Improves the productivity and reduces the lead-time
2) Reduces design time of tool and fixtures by 12–25%
3) Offers higher quality designs which lead to better products
4) Facilitates higher quality documentation packages
5) Eliminates tedious and repetitive tasks
6) Offers savings in time and reduces the cost of introducing new parts
7) Provides opportunities for growth for design and drafting personnel
8) Offers better standardisation arrangements
9) Improves interface between design and manufacturing
10) Decreases response time for bids

Benefits by CAM

CAM:

1) Increase the productivity
2) Offers better and improved product quality
3) Facilitates better communication
4) Reduces prototype construction costs
5) Provides faster response to customers
6) Reduces floor space requirements
7) Improves the safety of the workforce by eliminating hazardous working conditions

Benefits by AS /RS

AS /RS:

1) Provides computer control of material, providing the status and location of material
2) for all subsystems within the CIM control architecture via a network
3) Reduces work related to inventory turnover time by 5%
4) Improves security and reduces product losses from handling damage and location errors
5) Eliminates costly repetitive handling and data management of material
6) Increases productivity and production flow
7) Reduces assembly completion time by 25%

Benefits by GT

GT:

1) Improves design and enhances standardisation of it
2) Reduces materials handling
3) Reduces production lead-time by 20–80%
4) Reduces work-in-progress by up to 80%
5) Reduces labour by 15–25%
6) Reduces tooling by 20–30%
7) Reduces setup time by 20–60%
8) Reduces order time and delivery by 15–130%
9) Simplifies production scheduling and cycle time
10) Improves quality control
11) Provides an ability to react quickly to changing market conditions, hence reduces delivery time for an order
12) Increases team work between company functions
13) Increases workforce and management morale
14) Reduces paperwork by 80%

15) Reduces scrap materials and rework by 15–75%
16) Allows better planning and increases labour productivity

Benefits by CAPP

CAPP:

1) Reduces new product design lead-time
2) Reduces new part introduction cost
3) Increases capacity utilisation
4) Reduces set up and tool design time
5) Reduces process planning time
6) Reduces scrap and perishable tooling
7) Rationalises and reduces the number of parts in database

Benefits by CAQC

CAQC:

1) Reduces inspection time and indirect cost of inspectors by 80–90%
2) Reduces quality related costs by 50–90%
3) Helps to improve the quality of the product by thorough checks
4) Reduces scrap and rework by an on-line quality control procedures
5) Improves working conditions

Benefits by CM

CM:

1) Reduces the throughput time by up to 30%
2) Reduction in work-in-process inventory by 30–70%
3) Improves labour savings by 20–30%
4) Reduces engineering design costs by 15–30%
5) Provides lower setup time
6) Reduces overall lead-time by 30–60%
7) Provides efficient material handling
8) Offers improved output quality by 200–300%
9) Improves productivity by 50–150%
10) Gives improved job satisfaction for operators
11) Simplifies planning and control procedures

Benefits by Robots

Robots:

1) Increases productivity
2) Improves and provides more consistent product quality
3) Reduces scrap and reworking costs
4) Reduces cost of quality control
5) Reduces raw goods inventory
6) Provides direct labour cost savings
7) Provides savings in related costs such as lighting, cooling and in safety related matters
8) Provides savings from correctly forecasting production schedules
9) Can be used in any unpleasant and hazardous tasks
10) Improves machine utilisation
11) Provides reduction in training time compared to new workers

Benefits by Automated Assembly

Automated Assembly:

1) Provides better product quality with higher levels of consistency and uniformity
2) Improves productivity hence decrease manufacturing costs
3) Provides improved safety in dangerous assembly environment
4) Reduces floor space needs
5) Offers efficient production scheduling
6) Offers flexible approach to materials handling and could handle hazardous materials safely
7) Reduces inventory costs and amount of human labour in materials handling
8) Provides higher level of production reliability and reduces number of rejected parts and can operate in hazardous environments
9) Offers compatibility with production and storage equipment
10) Reduces warranty costs due to higher levels of quality.

Benefits of FMS

FMS:

1) Increase flexibility within a family of parts
2) Offers random feeding of parts and simultaneous production of different parts
3) Provides an ability to handle different materials
4) Provides an ability to continue production even if one machine breaks down
5) Improves use of manpower, equipment, and facilities and offers better management control

6) Decrease setup time and lead time by 40%
7) Decreases production costs by 50%
8) Reduces tooling costs by 30%
9) Reduces labour requirements by 50–90%
10) Provides efficient machine usage
11) Reduces work-in-process inventory by 9%
12) Improves product quality
13) Reduces rework and scrap
14) Reduces production floor space requirements by 42%
15) Increases productivity by 200–350%

Benefits of CIM

CIM:

1) Increases gain in engineering productivity by 300–350%
2) Increases gain in capital equipment operating time by 200–300%
3) Reduces work in progress inventory by 30–60%
4) Reduces material costs by 13–30%
5) Reduces production costs by 14–30%
6) Reduces design costs by 15–30%
7) Improves operating profit by 112–310%
8) Reduces tendering time by 80–90%
9) Reduces delivery time by 50–73%
10) Reduces overall lead-time by 30–60%
11) Improves quality and reduces scrap by 200–500%

A 2.2 THE KNOWLEDGE-BASE OF DSSCIM

The knowledge base gathered from literature, which is a summary of the information contained in Section A 2.1, is given in Tables A 2.1 and A 2.2 (pages 154 and 155) for the attributes discussed in Chapters four to six. The percentage values in Tables A 2.1 and A 2.2 are utilised in the DSSCIM to derive results shown in Chapter seven. However, the decision-makers are given a facility to change the values, which they regard as inappropriate according to their knowledge or do not satisfy the concerns of the company.

Table A 2.1 Attributes and their improvement ratios (for OA to AS/RS)

	Attributes offered in the strategic decision profile	OA	CAD	CAM	CAPP	CAQC	AS/RS
DCR	Minimise Direct Labour	10	10				30
	Reduce Material Costs			30			
	Reduce Material Handling Costs						70
	Reduce Tooling/Fixture Costs			60			
	Reduce Scrap					70	
ICR	Reduce Maintenance Costs			30			
	Reduce Prod. Plan. & Control				70		
	Reduce Floor Space	15					
	Reduce Inspection					**80**	
PIPF	Increase Manufacturing Productivity		40	40			
	Improve Machine Utilisation				70		
	Reduce Work In Progress			60	60		25
	Reduce Rework			70		80	
	Reduce Lead-Time	10	20	60			
PIHF	Reduce Paperwork	40	40				
	Improved Working Conditions	10	20	15	15	15	70
	Increase Design Flexibility		**18**	25			
	Increase Production Flexibility		10	45			
BS	Increase Product Growth		15				
	Increase Competitiveness				10		
	Ability to Introduce New Products		20	45			
	Ability of the System to Expand						40

Table A 2.2 Attributes and their improvement ratios (for Robots to CIM)

	Attributes offered in the strategic decision profile	ROBOTS	GT	CM	FMS	CIM
DCR	Minimise Direct Labour	70	10	**25**	**65**	50
	Reduce Material Costs					**14**
	Reduce Material Handling Costs	*30*				**20**
	Reduce Tooling/Fixture Costs		**25**	**20**	**30**	50
	Reduce Scrap	*70*				**35**
ICR	Reduce Maintenance Costs					**20**
	Reduce Prod. Plan. & Control					50
	Reduce Floor Space	45		30	**42**	50
	Reduce Inspection	*20*				50
PIPF	Increase Manufacturing Productivity	*65*		40		**200**
	Improve Machine Utilisation	80	30			50
	Reduce Work In Progress		**55**	**20**	**80**	**45**
	Reduce Rework	75	**50**			50
	Reduce Lead-Time	40	50	45	**60**	**45**
PIHF	Reduce Paperwork					50
	Improved Working Conditions	80	10	10	20	50
	Increase Design Flexibility					50
	Increase Production Flexibility				60	50
BS	Increase Product Growth					**60**
	Increase Competitiveness	*60*	**60**		*70*	**60**
	Ability to Introduce New Products				60	50
	Ability of the System to Expand				50	50

Note:

1. Attributes which have a value assigned on this table are associated with the particular AMT for impact-evaluation and quantification of intangibles.
2. Values, which are typed with bold font, are directly obtained from literature, and others are estimated during the research.

Index